Donald Emerson

INCENDIARY FIRES

Executive Editor: Jim Yvorra
Production Editor/Text Designer: Paula Huber
Art Director: Don Sellers
Assistant Art Director: Bernard Vervin

Typesetting & Pasteup: Prestige Editorial and Graphics Services, Inc.
Washington, DC
Printer: R.R. Donnelley & Sons Company
Harrisonburg, VA
Typeface: Times Roman

INCENDIARY FIRES

A Reference Manual for Fire Investigators

Steven E. Swab

Robert J. Brady Company • Bowie, MD 20715
A Prentice-Hall Publishing and Communications Company

Incendiary Fires: A Reference Manual for Fire Investigators

Library of Congress Cataloging in Publication Data

Swab, S. E. (Steven E.), 1941–
Incendiary fires.

Includes index.
1. Fire investigation—Handbooks, manuals, etc.
I. Title.
TH9180.S94 1983 363.2'562 82-25313

ISBN 0-89303-540-8

Prentice-Hall International, Inc., London
Prentice-Hall Canada, Inc., Scarborough, Ontario
Prentice-Hall of Australia, Pty., Ltd., Sydney
Prentice-Hall of India Private Limited, New Delhi
Prentice-Hall of Japan, Inc., Tokyo
Prentice-Hall of Southeast Asia Pte. Ltd., Singapore
Whitehall Books, Limited, Petone, New Zealand
Editora Prentice-Hall Do Brasil LTDA., Rio de Janeiro

Printed in the United States of America

83 84 85 86 87 88 89 90 91 92 93 10 9 8 7 6 5 4 3 2 1

CONTENTS

DISCLAIMER

TABLES

SCHEMATICS

PREFACE

Most fire investigation seminars never address the issue of what "real-world" arsonists are doing. Also, there is limited information available beyond the point of discussing flammable liquid patterns, beading on wires, or alligatoring effects.

We must recognize that people (arsonists in particular) have intelligence and imagination. Not everyone pours flammable liquid throughout a house and lights it with a flare.

This latter statement should have special meaning, considering the fact that presently there is a nationwide effort toward public awareness of the arson problem. That effort is known also by professional 'torches,' so we must presume that they will change tactics from flammable liquids to other methods that look accidental. As will be seen herein, these alternative methods need not be sophisticated. All they need to achieve is the "accidental" look.

We will not be concerned with admissions, confessions, motives, laws, courtrooms, or any of the other various aspects of the criminal side of arson investigation. The purpose of this text is to demonstrate the "before," "how-it-is-done" and the "after." The "after" also will include analysis of the remains, whenever possible.

All of the methods described here will present the appearance of an accidental fire, provided the fire has sufficient time to burn and destroy certain evidence. Careful scene analysis should provide clues that will remove the doubt of whether a fire was accidental.

Careful analysis does not include tramping over, around, and through a fire scene, sticking the probe of a hydrocarbon detector (sniffer) into every place it is presumed, or hoped, the residue of a flammable liquid may remain. Careful analysis means, among other things, a thorough, methodical, meticulous layering downward through the debris to locate the point of origin and source of ignition. It is a dirty, time-consuming task that *must* be accomplished if one is to remain unchallenged in a court of law.

It should be noted that the purpose of this manual is not to instruct one in how to set a fire. Neither is the purpose to train analysts and investigators how to process a fire scene. It is presumed that anyone referring to this manual has been thoroughly trained in the proper methods of scene processing, layering techniques, and evidence recovery and sampling techniques.

The author strongly recommends training (beyond the scope of this manual) in the areas of electronics (how radios, tv sets and the like work);

heating, cooking, and cooling appliances (repair and operation); electricity (commercial and residential); and the theory of combustibility, flame spread, and heat transfer of materials commonly encountered in everyday situations and environments.

Professional torches study these items closely, many times by going to night school at a nearby technical school, and so gain knowledge on how things function, why they function in certain ways, and what to expect as a result of both normal and abnormal functioning. It only follows, then, that fire analysts also must take time to obtain a thorough knowledge in these areas.

Additionally, the following are a few of the many text references the author recommends to analysts and investigators. The partial list includes: *National Fire Codes* (NFPA); *Life Safety Code* (NFPA); *Basic Building Code* (BOCA); *Fire Protection Handbook* (NFPA); *American Electrician's Handbook,* Carr (McGraw-Hill); *Mark's Standard Handbook for Mechanical Engineers,* Baumeister (McGraw-Hill); *Handbook of Engineering Fundamentals,* Eshbach (Wiley Publishing); *Modern Refrigeration and Air Conditioning,* Althouse (Goodheart-Wilcox Pub); *Gas Engineers Handbook* (Industrial Press); and *Domestic and Commercial Oil Burners,* Burkhardt (McGraw-Hill).

Be aware that many devices in this manual can be used in combination with other devices.

Steven E. Swab
President, ICO, Inc.
Fire Cause and Origin Consultants
Boca Raton, FL 33433
(305) 487-2435

1

INCENDIARY REFERENCE GUIDE

Incendiary Reference Guide

Primary Evidence Uncovered	Secondary Evidence (Some or all of which may, or may not, be located)	Ignition Method to Consider
Aluminum Powder	Sodium Peroxide, Sodium Nitrate, Potassium Nitrate, Barium Nitrate, Sodium Dichromate, Potassium Dichromate, Potassium Permanganate	Igniter #2, Incendiary #3
Artificial Fireplace Logs	Ordinary Combustibles	Chemical Method #9
Balloons	Sulphuric Acid, Flammable Liquids, Ordinary Combustibles	Delay #2
Barium Nitrate	Aluminum Powder, Magnesium Powder, Sodium Nitrate, Sodium Dichromate, Potassium Dichromate, Potassium Permanganate	Incendiary #3
Benzene	White Phosphorus, Carbon Disulfide, Yellow Phosphorus, Copper Sulfate	Chemical Method #2
Bi-Metallic Strips	Battery, Wires	*See* Electronics
Brown Powder	Powder (may have copper tint)	Chemical Method #9
Burglar Alarm Systems		*See* Electronics
Burnt Sugar Odor		Chemical Method #6
Candle Wax	Flammable Liquids	Chemical Method #7, Delay #4
Carbon Disulfide	White Phosphorus, Benzene, Yellow Phosphorus, Copper Sulfate	Chemical Method #2
Carmel Candy Odor		Chemical Method #6
Cartridge Fuses		*See* Electronics
Charcoal	Flammable Liquids	Incendiary #4
Charcoal, Powdered		Igniter #4
Chlorine (HTH)	Brake Fluid, Glycerin, Hair Creme	Chemical Method #3
Chocolate Fudge-like Substance	Concentrated Sulphuric Acid, Time Fuses	Igniter #1
Cigarette Butts	Match Heads, Ordinary Combustibles, Match Packs	Delay #1
Circuit Breakers		*See* Electronics
Cod Liver Oil	Rags	Chemical Method #4

Incendiary Reference Guide (cont'd.)

Primary Evidence Uncovered	Secondary Evidence (Some or all of which may, or may not, be located)	Ignition Method to Consider
Concentrated Nitric Acid	Match Heads, Concentrated Sulphuric Acid	Igniter #3
Egg (sour and putrid) odor	Most Probably "Alert 80" Odorant used in Natural and LP Cases	Consider all Heating, Cooking and Cooling Appliance Ignition Methods
Electrical Charcoal Starter		*See* Appliances
Flashlight	Batteries, Photocell, Relay	*See* Electronics
Flex Tubing (Natural Gas)	Natural Gas Heating Appliances	*See* Appliances
Fuming Nitric Acid	Turpentine, Other Flammable Liquids	Igniter #5
Furnaces		*See* Appliances
Fuses, Edison Base		*See* Electronics
Gunpowder	Flammable Solvents, Paper, Rags, Flake Aluminum Thermite, Ordinary Combustibles	Igniter #4
Hemp Oil	Rags	Chemical Method #4
HTH (Clorine)	Glycerin, Brake Fluid	Chemical Method #3
Hot Plates	Flammable Liquids	*See* Appliances
Hygrometric Triggering		*See* Electronics
Jelly-like Substance	Soap Powder, Flammable Liquids	Incendiary #1, Incendiary #2
Kerosene Odor	Wax Residue	Chemical Method #8
Light Bulbs		*See* Electronics
Limit Switches		*See* Appliances
Linseed Oil	Rags	Chemical Method #4
Lysol Odor	Chlorine, Glycerin, Brake Fluid	Chemical Method #3
Match Heads	Fuses, Concentrated Sulphuric Acid, Concentrated Nitric Acid	Igniter #3
Metallic Potassium	Milky Residue on Debris	Chemical Method #1
Milky Residue on Debris	Metallic Potassium	Chemical Method #1
Model Rocket Fuse		Delay #3

Incendiary Reference Guide (cont'd.)

Primary Evidence Uncovered	Secondary Evidence (Some or all of which may, or may not, be located)	Ignition Method to Consider
Model Rocket Igniters	Batteries, Delay Timers	See Electrical Section
Natural Gas		*See* Appliances
Nichrome Wire		*See* Electronics
Noise-Activated Trigger	Battery, Circuit Board	*See* Electronics
Offensive Odor (may cause watering of eyes)	Odor Most Probably that of Propane Gas	Consider all Heating, Cooking and Cooling Methods
Oil Pumps		*See* Appliances
Ovens		*See* Appliances
Paint (Oil from top of can)	Charcoal, Ordinary Combustibles	Incendiary #4
Photocell	Battery, Transistor Relay, SCR	*See* Electronics
Poppyseed Oil	Rags	Chemical Method #4
Potassium Chlorate	Sugar, Sulphuric Acid	Chemical Method #6, Igniter #1
Potassium Dichromate	Aluminum Powder	Incendiary #3
Potassium Nitrate		Igniter #4, Incendiary #3, Delay #3
Potassium Permanganate	Glycerin or Hair Creme with Glycerin, Aluminum Powder	Chemical Method #5, Incendiary #3
Prophylactics	Sulphuric Acid, Flammable Liquids, Ordinary Combustibles	Delay #2
Proximity Trigger		*See* Electronics
Radio Transmitters		*See* Electronics
Refrigerators	Flammable Liquids	*See* Appliances
Rotten Egg Odor	Sulphur, Gunpowder	Igniter #4
Rotten Meat and Cabbage Odor	Most Probably a Sulfide such as BP Captan Odorant in Natural and LP Gases	*See* Appliances
Sawdust	Flammable Liquids, Wax	Incendiary #2
Silicone Controlled Relay	Photocell, Transistors, Batteries	*See* Electronics

Incendiary Reference Guide (cont'd.)

Primary Evidence Uncovered	Secondary Evidence (Some or all of which may, or may not, be located)	Ignition Method to Consider
Soap Powder	Flammable Liquids	Incendiary #1
Sodium Dichromate	Aluminum Powder	Incendiary #3
Sodium Nitrate	Aluminum Powder, Potassium Nitrate, Barium Nitrate, Sodium Dichromate, Potassium Dichromate, Potassium Permanganate	Incendiary #3, Delay #3
Sodium Peroxide	Aluminum Powder	Igniter #2
Sour, Meaty and Onion Odor	Odor most Probably that of Mercaptan Used as Odorant in Natural and LP Gases	Check all Heating, Cooking and Cooling Ignition Methods
Sulphur, Ground		Igniter #4
Sulphuric Acid	Potassium Chlorate, Sugar, Match Heads, Concentrated Nitric Acid, Balloons	Chemical Method #6, Igniter #3, Delay #2
Sweet, Pleasant Odor	Odor Could be Methane, as in Natural Gas, or Melted Polyurethane from Furniture	*Careful* Analysis is Necessary
Television Sets		*See* Electronics
Timer Circuits		*See* Electronics
Touch Switches		*See* Electronics
Tung Oil	Rags	Chemical Method #4
Turpentine	Fuming Nitric Acid	Igniter #5
Voice-Activated Trigger	Batteries, Circuit Boards	*See* Electronics
Wax Residue	Kerosene Odor, Flammable Liquids, Sawdust	Chemical Methods #7 and #9, Incendiary #2, Delay #4
White Phosphorus	Carbon Disulfide, Benzene, Yellow Phosphorus, Copper Sulfate	Chemical Method #2
Yellow Phosphorus	White Phosphorus, Carbon Disulfide, Benzene, Copper Sulfate	Chemical Method #2

2

CHEMICALS

INTRODUCTION

The successful use of chemicals to ignite a fire depends on the fact that certain substances will react with one another when they are brought in contact.

The exothermic (liberating of heat) effect causes the open burning, provided that one of the chemicals is an oxidizer (gives off oxygen), which, in effect, "feeds" the fire.

From an evidence procurement viewpoint, investigation of chemically ignited fires is extremely difficult. In most instances the evidence will be consumed by the fire, provided the fire ignited successfully and burned with sufficient intensity for a long enough period of time.

Chemical Method #1

Metallic potassium will ignite when combined with water, so it becomes an excellent incendiary device for use in warehouses and other similar areas in which the arsonist has time to set up the device without being seen.

A certain quantity is placed under a hole in a roof immediately prior to a storm. Quantities placed under several holes aid in the certainty of ignition and would most probably assure a major conflagration. Care is usually exercised to insure that multiple origins will not be detected. The placement of the chemical on top of highly combustible materials, such as cardboard boxes and crates, is obviously desirable.

Moisture (humidity) is the key factor needed for ignition, although ambient temperature is equally as important.

Once a temperature of approximately 70° Fahrenheit is reached, the metallic potassium may ignite and this alternative controlling factor makes this ignition method somewhat hazardous.

From the analytical aspect, if any residue remains, it will probably look milky. If, after examining the fire scene, there is no logical explanation for an ignition source, and the fire occurred during a high humidity time, consider this method of ignition in your final analysis.

Chemical Method #2

As with Chemical Method #1, all of the chemicals needed for this method are available across the counter at chemical supply houses. These chemicals are white phosphorus and carbon disulfide. If carbon disulfide is not available, benzene can be substituted. Yellow phosphorus will work also, but not red phosphorus.

As shown in Table 2-1, white phosphorus is used in rat poisons and other medicines. A jar labeled "Rat Poison" would not be unusual in the cellar or basement of a structure, so it represents a logical storage container without being obvious. A basement is also a logical place to set up this method. (As will be seen in subsequent methods, proper placement of the ignition device is crucial to avoid early detection.)

With Chemical Method #2, not only will the fire start low in the structure, but the heating ducts will help to spread the fire.

The ductwork is favorable as the surfaces reflect heat inward and, simultaneously, provide a means of drafting and heat. Spontaneous ignition usually occurs at or above 85° Fahrenheit.

It must be noted that this is a very hazardous incendiary. Should you encounter it, consider the arsonist to be a professional with experience in handling white phosphorus. (Copper sulfate solution is usually available to clean the skin in the event of a spill, so investigators should also be aware of the meaning of the presence of this material after a fire or at a suspect's house.)

Normally, the "mix" is 200 cc of the substances, which is achieved by pouring approximately 90 cc, followed by the insertion (with tweezers) of the white phosphorus sticks into the bottle to make the desired quantity. The top is then secured (no rubber stoppers or sealers can be used), and the mixture is shaken for about one and a half minutes to dissolve the white phosphorus completely.

As long as the container is kept securely sealed the mixture will keep indefinitely. Tests conducted by the author indicate it takes about 1½ minutes to ignite, at 80° Fahrenheit.

Application of the chemicals can take several forms. One method is to presoak the area of flooring above the floor registers of the forced air heating duct system. The presoak can be any flammable liquid such as a mixture of gasoline and fuel oil.

After turning the thermostat "down," and allowing the ductwork to cool off, the mixed chemicals are poured into the ductwork. After dumping the mixture into the ductwork, one simply turns the thermostat "up" to demand heat which, in a forced air system, also provides an air flow. Resulting evaporation causes ignition of the white phosphorus, which in turn "blows" out of the floor registers and ignites the combustibles or flammable presoaks immediately around the floor register.

This chemical fire will not burn extremely hot, but it is persistent and burns with a dense, white, irritating smoke. Reports of this phenomenon from firefighters should be noted.

Another reference note is that this method can "flame-up" a second time if initial extinguishment was by water. The second flame-up will occur after the water evaporates.

The following photographs represent a series of tests using this method. For these tests the mixture was poured down a wooden wall near a forced air diffuser. Ignition, as shown in Figure 2-1, occurred in approximately 1½ minutes.

In an effort to determine what residue would remain, an overhaul procedure was effected that yielded some very interesting information for fire analysts.

Evidence indicates the possibility of phosphoric acid eventually being found, along with the fact that this mixture could burn hot enough to burn through aluminum ductwork. The residual could also look like a white-red paste; however, our tests did not bear this out.

A second item noticed was the presence of smoke with re-kindling after evaporation of the water used for the suppression process (see Fig. 2-3).

Another visual sign was that each time the "Mini-Quick" bar struck the wood, sparks would be created similar to the sparks when a piece of flint is struck in a cigarette lighter.

Any time suppression personnel report an indescribable odor and reports are received of sparks coming from wood when struck by a metal bar, the investigator should immediately try to locate the residues previously mentioned.

Figure 2-1. Ignition 1½ minutes after application over the wall. It should be noted that the fire immediately "got away" inside the wall.

Figure 2-2. Close-up of the grill showing damage as a result of a burn-time less than five minutes.

Figure 2-3. Note smoke rising from the top of the grill/diffuser.

Chemical Method #3

As with other methods described here, this method utilizes "on-scene" materials. From the swimming pool, powdered HTH; and from the garage, a can of brake fluid.

A quantity of the HTH is poured over the combustibles to be ignited as shown in Figure 2-4.

The brake fluid is then poured (puddled) on top of the HTH. Self-ignition follows within a few minutes and usually results in a relatively high hot flame (see Fig. 2-5).

The brake fluid is immediately consumed. Remaining residue, if any, will be a white mushy material with a very distinctive chlorine odor, especially when a piece of the substance is broken away from the main "pile" (see Fig. 2-6).

Analysis may reveal the presence of HTH (chlorine odor) or the purple to black crystals if potassium permanganate was used. Fire suppression personnel need to be taught to preserve the scene if residues like these are to be found. This preservation includes not only holding the scene after suppression, but insistance upon careful application of water when attacking the fire. To preserve this type of evidence, the application of water should be in a fog pattern, not a straight-stream pattern as the latter tends to blow away the evidence.

Figure 2-4.

Figure 2-5.

Chemical Method #4

Chemical Method #4 is quite often encountered as a genuine accidental fire, probably because it is difficult to prove intent. The plan consists of nothing more than linseed oil-soaked rags lying on top of each other and, many times, inside an ordinary combustible such as a cardboard box. Ultimately, the chemically reactive materials, such as the unsaturated vegetable oils, are absorbed in the rags, which then self-heat and self-ignite.

There are several oils that could be used for this purpose. The most reactive to least reactive seem to be linseed, tung, hemp, poppyseed, and cod liver oils, respectively.

Ignition times will vary depending on the oils and rag materials used. Success of the method also includes consideration of the types and quantities of additional materials to be ignited. The time frame for ignition can vary from a few hours to several days.

Chemical Method #5

This method can be used to ignite numerous items, but most often is used as a delay device for ignition of a floor presoaked with a flammable liquid and as an ignition source of clothing in clothing stores.

The chemicals involved are potassium permanganate (available across the counter) and any hand lotion or hair creme that contains glycerin. An alternate for potassium permanganate is powdered HTH.

When used as a sabotage device in clothing stores, an envelope (usually 3½ by 6½ inches) is employed. Starting in one corner of the envelope the lotion (or creme) is squeezed out along the bottom (see Fig. 2-7).

This lower section of the envelope is then folded over and the potassium permanganate is poured into the next "section."

Keeping the envelope folded so the chemicals don't come in contact with one another, the flap is sealed. At the scene, the envelope is unfolded, shaken

Figure 2-6.

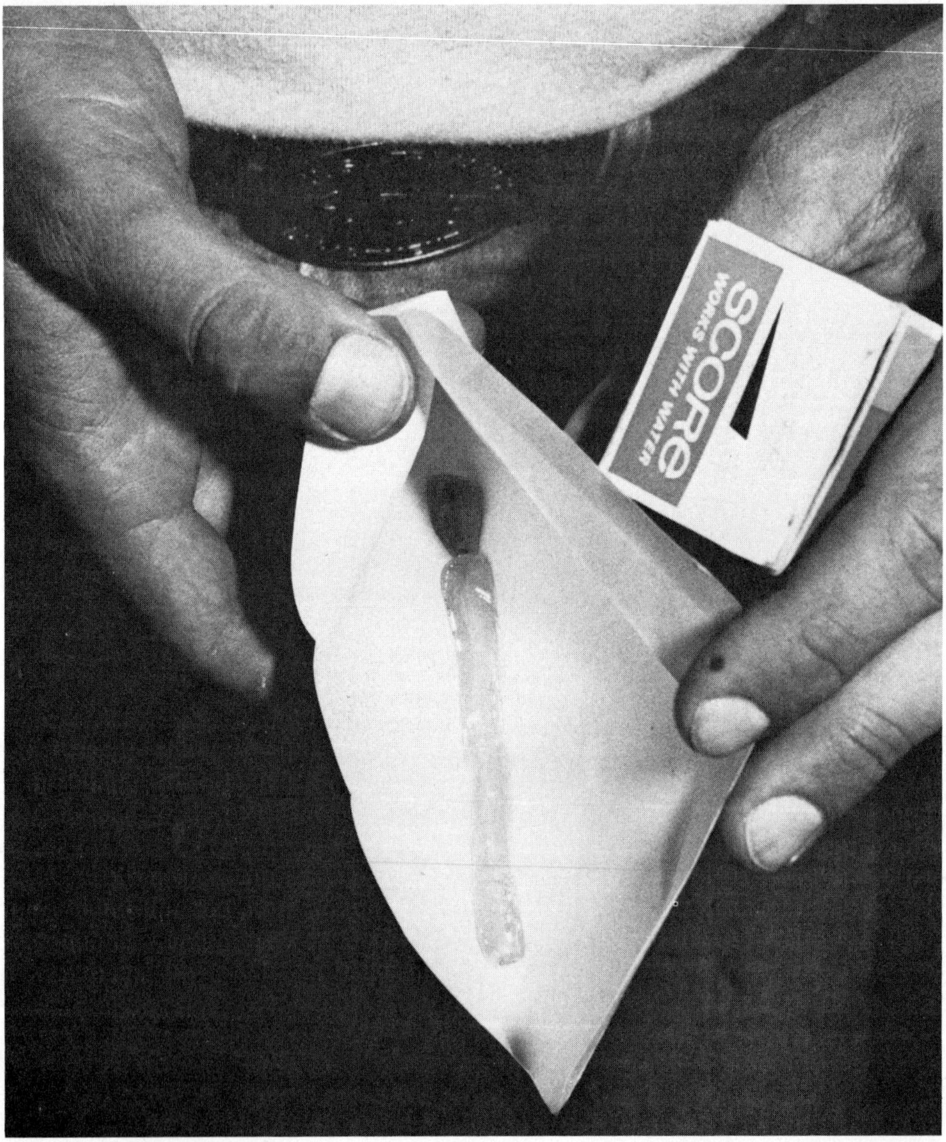

Figure 2-7.

(to mix the materials), and placed inside a rack of clothing or on the combustibles to be ignited (furniture, boxes, or any other flammable combustibles).

As shown in Figure 2-8, the mixture will first start to smoke and very shortly thereafter rupture into open flame, as shown in Figure 2-9.

Chemical Method #6

Utilization of this method usually occurs in apartment building storage rooms and warehouses containing cardboard boxes or similar combustibles.

Figure 2-8.

Figure 2-9.

Usually about three teaspoons of potassium chlorate are mixed with one teaspoon of sugar. This mixture is then poured into a small cardboard box as shown in Figure 2-10.

Figure 2-10.

Two teaspoons of sulphuric acid are then poured into a test tube or other glass container, which is then sealed with masking tape.

The container with the sulphuric acid is then laid on its side in the cardboard box and the acid eats through the masking tape (usually between 60 and 120 minutes). (For test purposes, a plastic glove over the bottle speeded up the process. See Fig. 2-11.)

Analysis after the burn indicates several items of interest to the fire analyst.

If any acid remains on the floor, laying paper on the acid will cause the paper to turn brown or black.

As shown in Figure 2-12, there also remains a black charred clump of crusty material with a carmel candy or burnt sugar odor. If this is found, it should be properly gathered, boxed, and sent to a laboratory with comments on what was found, how it looked and smelled, and what the analyst suspects.

Chemical Method #7

This ignition method, although not really a "chemical" one, uses products not associated with any other section in this text, so it is included here.

The area is first presoaked with a low volatility fuel such as fuel oil or kerosene. (Restaurants often use "degreasers" that contain a kerosene base.)

A candle is placed in a small container by whatever means will insure that the candle will not fall over. After the candle is secure, rags are placed around and against the candle with "trailers" extending a few inches away from the container. The trailers are presoaked with the same flammable liquid and laid on top of the presoaked floor. (The candle can also be stuck on a nail rather than working to affix it to a container.)

By marking off ½- or 1-inch intervals on the candle, the delay time can be projected. Testing similar candles will provide data on how long it takes to burn off one inch. Shielding from air currents is also a requirement, so note air flow potential during analysis.

Using highly volatile flammable liquids, such as gasoline, is not to be expected as the vapors may ignite prematurely. Analysis of the debris may yield the presence of wax and the material used as the wick. Usually this method is very thorough and nothing remains that is conclusive.

Chemical Method #8

Firecubes

As shown in Figure 2-13, these cubes consist of a solidifying agent and a flammable liquid such as kerosene.

Typically, they come in a package of 60 cubes per box and are sold in almost any hardware or food store for use in igniting barbecues, fireplaces, and campfires.

Ignition is by open flame devices such as matches or by electronic ignition devices producing open flames, such as described in Chapter 4.

The cubes do tend to burn persistently, which is desirable; however, they also give off a great amount of black smoke that could possibly be detected,

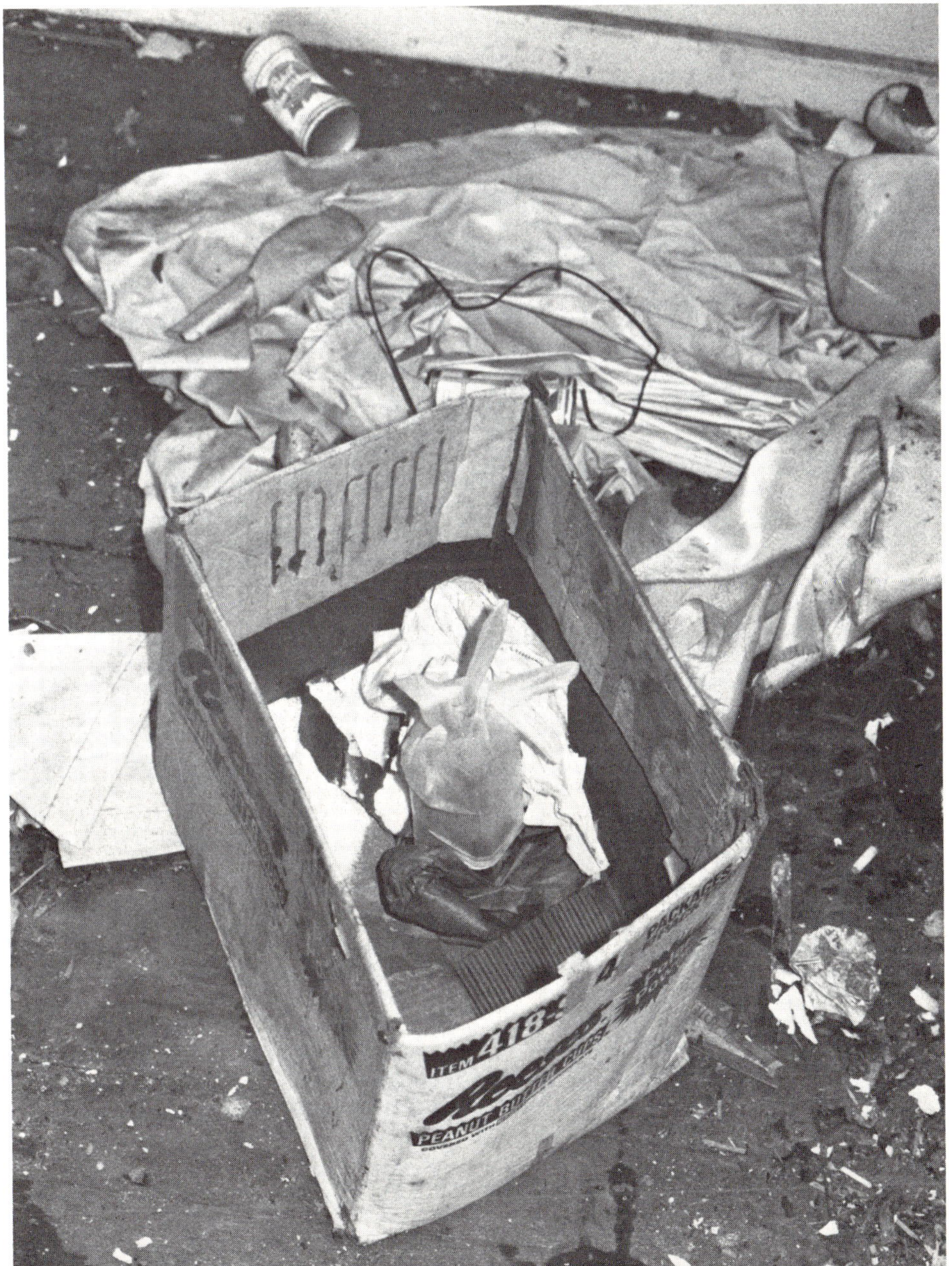

Figure 2-11.

Figure 2-12.

Figure 2-13.

depending on where the cubes were placed and what they were intended to ignite. While stored in wooden kitchen cabinets, they could be ignited "accidentally" or they could also be used against fabric-covered furniture. Depending on how successful the fire was, there may or may not be remnants of a wax residue with a distinctive odor of kerosene.

Chemical Method #9

Artificial Logs

As shown in Figure 2-14, these "logs" come in packages and are composed of sawdust, wax, and a copper-based coloring additive.

When used as intended, they give off approximately 50,000 BTUs per unit over a three-hour period. When used as an incendiary device, however, the logs are usually broken open, spread around in small piles, and ignited by any open flame device.

As with other devices, residue may remain after suppression of the fire. This residue could look like a brown or copper-tinted powder. On a test burn on porous patio stones (see Fig. 2-15) the residue was black. Initially, after burning, the residue did have a slight tint of copper to it; however, layering down to the bottom revealed only the black residue. The residue had a "waxy" feel to it with no distinctive odor.

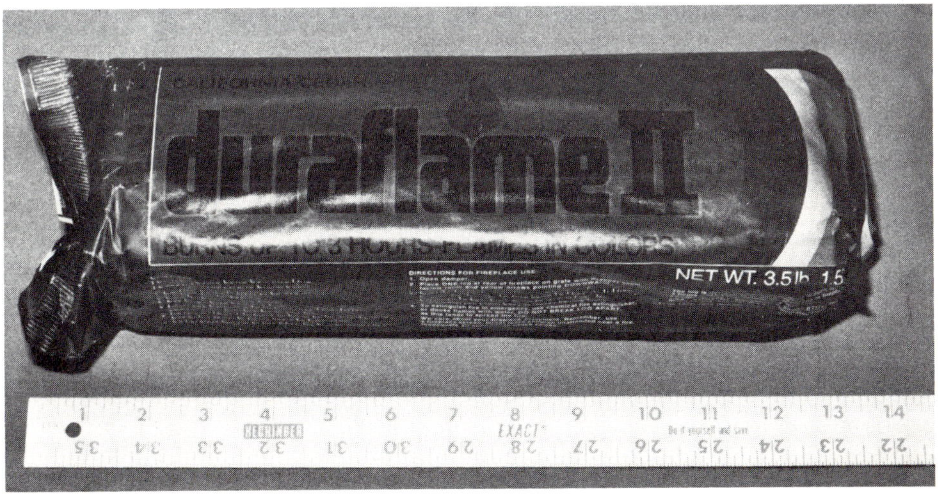

Figure 2-14.

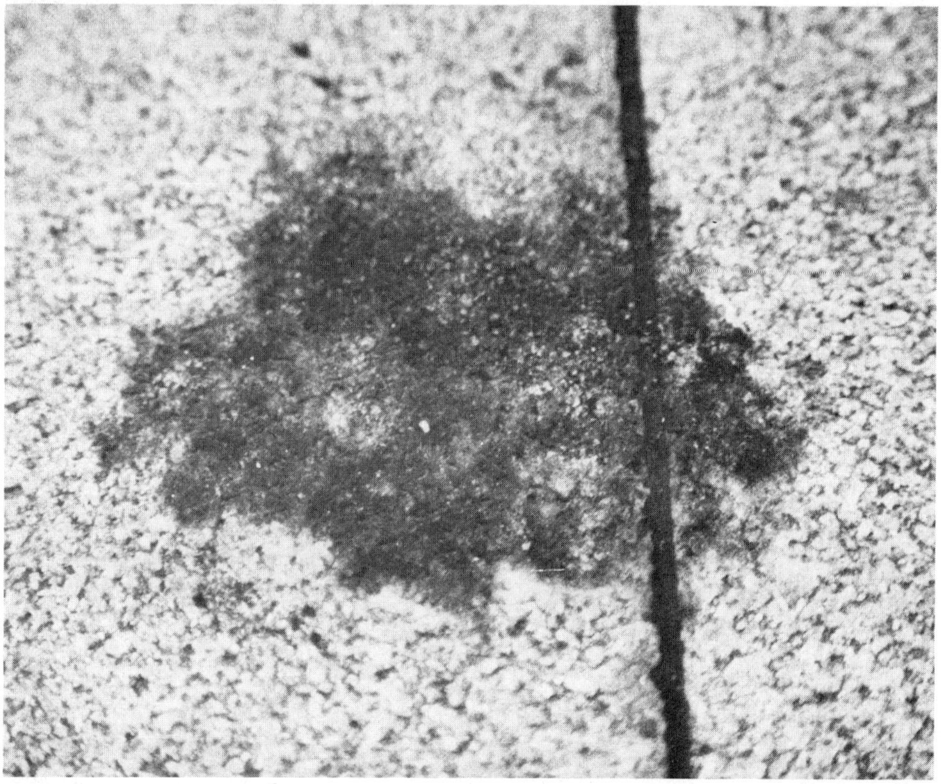

Figure 2-15.

Table 2-1: Chemical Names and Substitutes

Chemicals	Substitute	Comments
Acetic Acid	Vinegar	
Aluminum Oxide	Aluma	
Aluminum Potassium Sulfate	Alum	
Aluminum Sulfate	Alum	
Ammonium Hydroxide	Ammonia	
Barium Nitrate	Nitrate of Baryta	Used in Fireworks
Benzene	Benzol, Coal Naptha	Used in Dry Cleaning
Black Iron Oxide	Black Rouge	Art & Hardware Stores
Carbon Carbonate	Chalk	
Carbon Tetrachloride	Perchloremethane	Used in Dry Cleaning
Calcium Carbide		Usually Found in Hardware Stores, Miners' Lamps for Illumination
Calcium Chloride	Weed Killer	Ice Thawing, Basement Dehumidifiers
Calcium Hypochloride	Bleaching Powder	Use Only Dry Powder
Calcium Oxide	Lime	
Calcium Sulphate	Plaster of Paris	
Carbon Disulfide	Carbon Bisulfide	Insecticides
Carbonic Acid	Seltzer	
Cupric Chloride	Etching Agent for Galvanizing	Found in Paint & Photo Supply Stores
Cupric Sulfate	Blue Stone	Photoengraving & Insecticides
Ferric Oxide	Iron Rust	
Flake Aluminum Powder	Silver Bronzing Powder	Art Supply Stores
Glucose	Corn Syrup	
Glycerin	Glycerol	Food & Cosmetics
Graphite	Black Lead	
Hydrochloric Acid	Muriatic Acid	
Hydrogen Peroxide	Peroxide	
Lead Acetate	Sugar of Lead	
Lead Tetroxide	Red Lead	
Magnesium Powder	Photoflash Powder	Also Found in Filings or Drillings in Machine Shops
Magnesium Silicate	Talc	
Magnesium Sulfate	Epsom Salts	
Naphthalene	Mothballs	
Nitric Acid, Fuming	Aqua Fortis, Azotic Acid	Electroplating Shops
Parafin Wax	Ceresen	Used in Candlemaking
Phenol	Carbolic Acid	
Potassium Bitartrate	Cream of Tartar	
Potassium Chlorate	Potassium Oxymuriate	Disinfectant
Potassium Chromium Sulfate	Chrome Alum	
Potassium Dichromate	Potassium Bichromate	Used in Leather Tanning, Etching
Potassium Nitrate	Saltpeter	
Potassium Permanganate	Wood Preservative, Deodorants, Medicines	
Red Iron Oxide	Jeweler's Rouge Crocus	Hardware Stores
Silver Nitrate	Lapis Custic	Silvering Glass Photography

Tablc 2-1: Chemical Names and Substitutes (cont'd.)

Chemicals	Substitute	Comments
Silicon Dioxide	Sand	
Sodium Bicarbonate	Baking Soda	
Sodium Borate	Borax	
Sodium Carbonate	Washing Soda	
Sodium Chlorate	Used in Weed Killers	
Sodium Chloride	Salt	
Sodium Hydroxide	Lye	
Sodium Nitrate	Chile Saltpeter, Nitratine	Drug & Hardware Stores
Sodium Peroxide	Sodium Dioxide, Bleaching	Chemical Supplies
Sodium Silicate	Water Glass	
Sodium Sulfate	Glauber's Salt	
Sodium Thiosulfate	Photographer's Hypo	
Sulfur	Brimstone	Insecticides
Sulfuric Acid	Battery Acid	
Sucrose	Cane Sugar	
Toluene	Toluol, Ink Solvent	Paint Stores
Turpentine	Oil or Spirits of Turpentine	Paint Stores
White Phosphorus	Rat Poison, Medicine	Found in Chemical Supply Houses
Zinc Chloride	Tinner's Fluid	
Zinc Dust	Spelter, Anticorrosive Paint	Paint Stores

3

IGNITERS, INCENDIARIES & DELAYS

INTRODUCTION

Generally, the devices in this chapter are considered "exotic" and not too likely to be used. However, since various "manuals" have been prepared on this subject, it is felt that such information should be included so that fire investigators will be able to formulate knowledgable opinions and conclusions about certain evidence.

Igniters are any materials that will start the burning process. The most common example is a match. It should also be noted that the most successful fires are ignited by open flame devices.

Incendiaries are any mixture of materials that will burn with sufficient heat and flame to insure hostile fire. Fuel oil, gasoline, and napalm are generally classified as incendiaries, even though most people call them accelerants. An incendiary also can be used against materials normally thought to be "non-burnable" such as metal, in which case the incendiary would be thermite.

Delay, as the name implies, provides the user with sufficient time to leave the scene and perhaps establish an alibi. At some predetermined time, the delay mechanism activates and ignites the fire.

Igniter #1

Fire Fudge

Material to be Ignited: Any incendiary listed in this chapter, except commercial termite. It is excellent for use against hay, straw, paper, and rags.

Ignition Sources: Time fuses, open flame, concentrated sulfuric acid.

Materials Needed: Granulated sugar, potassium chlorate.

Fire fudge is prepared by pouring 25 cc of clean water into a pan, warming it, and then adding 25 g of sugar. The mixture is stirred until dissolved, then covered and brought to a boil for a few minutes. At this point, a thermometer indicates a temperature of 100° C.

Immediately upon reaching this temperature, the pan is removed from the heat source and 50 g of potassium chlorate are added. After mixing the substances for about a minute, the mixture is poured into molds or a container. The result is a white fudge-like substance which, if kept in a tight container, will keep for quite some time.

Igniter #2

Aluminum Powder-Sodium Peroxide

Material to be Ignited: Napalm, white phosphorus solutions, flake aluminum thermite, thermate, paints, lacquers, and paraffin sawdust mixtures.

Ignition Sources: Water, fuses, diluted acids, open flame devices.

Materials Needed: Fresh sodium peroxide, aluminum powder or filings.

This igniter is prepared by filling a small can or jar approximately ¼ full of aluminum powder. The powder should be the finest (vs. coarsest) powder obtainable.

Using sodium peroxide (preferably granulated sugar consistency), enough sodium peroxide is added to fill the container (on top of the aluminum powder) half full. The container is then covered with the lid and shaken to mix the components. In a tightly closed container, it will keep for several weeks. A reaction will take place that causes ignition if the mixture comes in contact with water.

The usual method of ignition is to suspend the mixture in some manner (usually inside a paper envelope) over a moist area and provide some means of having the mixture and moisture come in contact with each other. A second way is to provide a place to spread out the mixture where it will come in contact with water (on top of boxes under the roof of an old warehouse).

Igniter #3

Match Head Igniter

Material to be Ignited: Flammable liquids, white phosphorus, thermate, paints and lacquers in open containers, paper, rags, excelsior and other "normally combustible" materials.

Ignition Sources: Fuses, open flame, concentrated sulphuric acid, and concentrated nitric acid.

Materials Needed: Any type of matches.

The preparation of this igniter is quite simple. Match heads are broken off or scraped off with a pocket knife.

Placed in a suitable container (wrapped in paper or tissue), the match heads are placed on the material to be ignited and the ignition source is activated. (See other areas of this text for combinations of ignition sources to be used as delays or remote devices.)

Igniter #4

Gunpowder

Material to be Ignited: Flammable solvents in open containers, flake aluminum thermite, thermate, paper, rags, and other combustibles.

Ignition Sources: Open flame, sparks, and fuses.

Materials Needed: Gunpowder as purchased from sporting goods stores and other suppliers. Gunpowder can be prepared from potassium nitrate, ground sulfur, and powdered charcoal, if necessary.

The usual application is "trailers" across and into the flammable to be ignited, with ignition of the gunpowder by one of the methods mentioned above.

Gunpowder is not affected by weather; however, it does burn rapidly, so the combustibles it is used against must ignite readily.

From an investigator's viewpoint, the trailer may remain as may residue from the gunpowder, which could create evidence if the fire failed to burn thoroughly.

Igniter #5

Turpentine-Fuming Nitric Acid

Material to be Ignited: Paper, napalm, and flammable liquids in open containers or in a "presoak" area of floors.

Ignition Sources: Adding turpentine to fuming nitric acid or vice versa.

Materials Needed: Turpentine and Fuming Nitric Acid.

When either of the materials is combined with, or comes in contact with, the other material, a flame is produced almost instantly.

The normal ignition is to utilize a "tipping" arrangement of containers where one is tipped over onto the other. The fuming nitric acid container will usually be made of glass.

The device is positioned such that the flame will impinge on the material(s) to be ignited (e.g., on top of a counter where wooden cabinets would ignite). Usually this device will not work well in cold weather or in a windy environment.

Incendiary #1

Napalm

Material to be Ignited: Paper, straw, hay, and wood.

Ignition Sources: Any igniter listed here, except white phosphorus or time fuses.

Materials Needed: Soap powder, gasoline, fuel oil, diesel oil kerosene, benzene or toluol (detergents will not work in place of soap).

The usual preparation is to use a double boiler and bring the water to a boil in the lower section to melt the soap chips in the upper section. Once this is accomplished, the flame/heat source is removed and the liquid is stirred into the melted soap mixture until it thickens into a paste with a consistency about equal to jelly.

Usually the mixture is spread over the materials to be ignited and subsequently ignited by a delay mechanism or device. Most manuals indicate that a delay over one hour will not be used as the gasoline may evaporate. In warm weather the delay will probably not exceed 15 minutes.

If this incendiary is used in a normally warm or hot room, the mixture will probably contain fuel oil rather than gasoline, whereas in cold weather (a time when napalm ignites readily), gasoline will probably be used.

Incendiary #2

Paraffin/Sawdust

Material to be Ignited: Wood, paper, rags, straw, and hay.

Ignition Sources: All igniters listed here, except time fuses, calcium hypochlorite or turpentine-fuming nitric acid.

Materials Needed: Sawdust, paraffin wax, beeswax or wax from melting candles.

As with Incendiary #1, a double boiler pan is desirable wherein the lower pan contains boiling water that melts the wax in the upper pan.

After melting the wax, the sawdust is added while stirring the mixture to insure consistency and equal parts mixture. The consistency should approach a jelly state, at which point the pots are removed from the heat source and the mixture is allowed to cool and harden.

Most often the incendiary is placed in a paper bag and positioned between cushions of furniture. The corner or edge of the bag is ignited. Since the bag and incendiary burn slowly, there is no risk involved in using a hand-held match.

When igniting wood, it is usually placed on a pile, such as on top of the ductwork or above a ceiling in a basement, whereby the subfloor is ignited. Residue may include the presence of wax with bits of sawdust or wood, depending on thoroughness of burn.

Incendiary #3

Thermate

Material to be Ignited: Metal, paper, rags, excelsior, and wood.

Ignition Sources: Virtually any igniter.

Materials Needed: Aluminum powder or magnesium powder in combination with any chemicals such as sodium nitrate, potassium nitrate, barium nitrate, sodium dichromate, potassium dichromate, or potassium permanganate.

The container used is first filled about one-third full with the metal powder, to which is added the oxidizer, to make a total volume of about two-thirds full. This combination is then rolled (not shaken) around to insure thorough mixing. The mixing process is usually done for a minimum of three minutes.

The preference is to use the mixture against metal or wood as it tends to burn hot but rapidly, usually too rapidly for combustibles such as paper, etc. It should also be noted that this incendiary is usually ignited by a delay device since the rapid ignition tends to burn anyone igniting it.

Investigators should also note that almost anyone using this device would usually practice on similar materials, so evidence to this effect may be found at a suspect's house. The device doesn't always work against the intended material, so practice on similar materials is usually intended to insure success when it is needed.

Incendiary #4

Paint

Material to be Ignited: Wood, paper, straw, and rags.

Ignition Sources: All igniters except phosphorus solutions or time fuses.

Materials Needed: Paint, lacquer, enamel or varnish except water-thinned paints, water emulsion paints, or prepared liquid shellac.

When this incendiary is used, it normally is not mixed prior to usage as the flammable part is the portion that floats on top of the paint can. The upper-portion liquid is poured directly onto the material to be ignited and then ignited.

A second method of application is to pour some of the liquid into a container (tin or plastic can) and, at the same time, pour some over pieces of charcoal arranged in a small stack around the base of the can. This entire device is set up on or near the bottom of crates or boxes and ignited. Likewise, this set-up could be placed around the base of the furniture and ignited. Charcoal provides a persistent heat with the open flame.

For the analyst, the presence of charcoal or a paint can could be indicative of a device, which should warrant more thorough investigation.

Delay #1

Cigarette Pickett

Material to be Ignited: Ordinary combustibles such as paper, straw, hay, tinder wood, fire fudge, and gunpowder.

Ignition Sources: Matches or open flame.

Materials Needed: Non-filter cigarette, kitchen or paper matches.

Cigarettes make reasonably good delay devices, provided they have sufficient oxygen and are placed such that the glow will not be detected prior to ignition.

The pickett delay is made by placing matches around the exterior of a cigarette as shown in Figure 3-1. Ignition occurs by carefully lighting the cigarette or by holding an open flame to the end of a cigarette. The match heads must be placed such that they will readily ignite the combustible material.

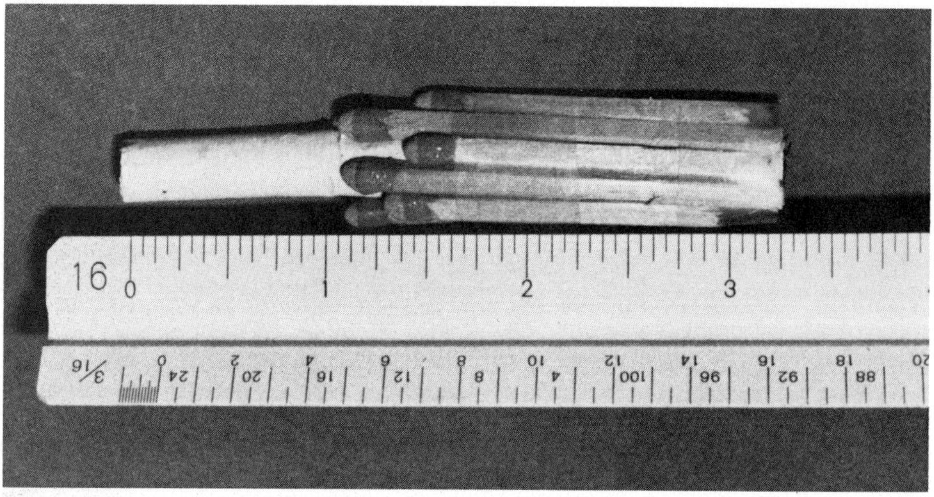

Figure 3-1.

A second method of using a cigarette delay is to insert the cigarette into a box of matches as shown in Figure 3-2. Again, it is important to place the device so the match heads will be sure to ignite the combustible. Depending on how much cigarette is exposed, the delay will be anywhere from three to twenty minutes.

The fire analyst must be alert for the presence of numerous unburned matches and the end of a cigarette inside some match sticks or inside the match box.

Delay #2

Rubber Diaphragm

Material to be Ignited: Fire fudge, match heads, aluminum powder.

Ignition Sources: Sulfuric Acid.

Materials Needed: Concentrated sulphuric acid, balloons or prophylactics, string, tape, and rubber bands.

The principle behind the success of this delay mechanism is that sulfuric acid will eat through the membrane within a certain period of time.

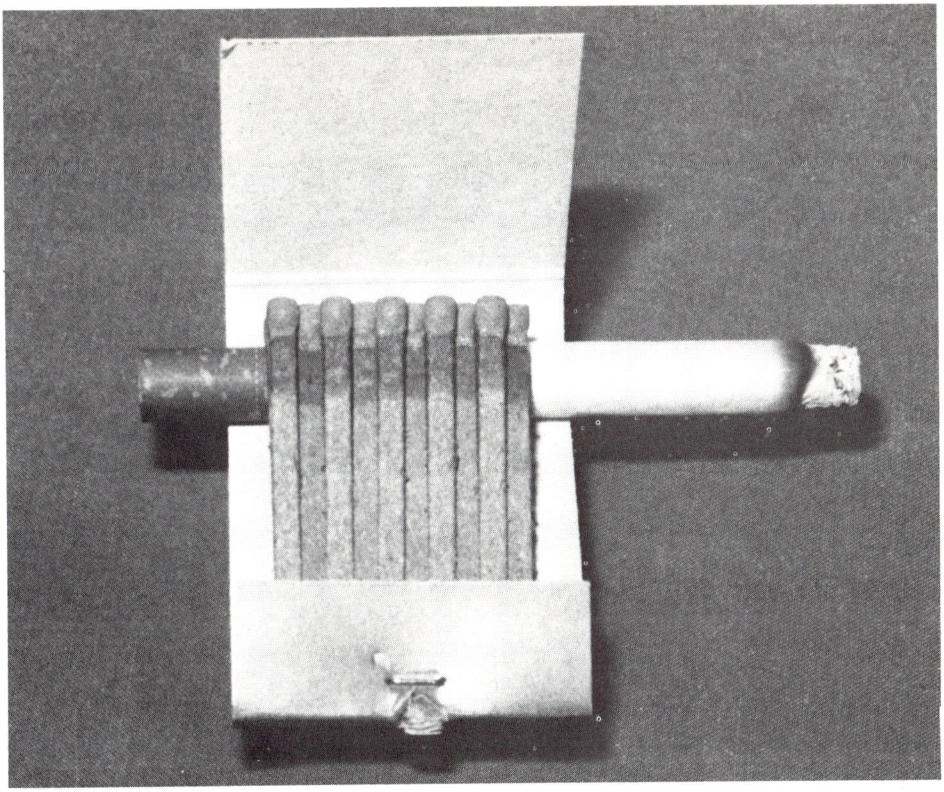

Figure 3-2.

This delay was used earlier in Chemical Method #6, where a plastic glove was placed over the bottle as shown in Figure 2-11.

String, tape, or rubber bands are used to fasten the membrane to the container. A prophylactic will usually provide a delay of 15 to 20 minutes. A plastic glove takes longer, usually 30 to 40 minutes.

Delay #3

String Fuse

Material to be Ignited: Sugar chlorate, fire fudge, match heads, and numerous other chemicals.

Ignition Sources: Open flame or match.

Materials Needed: Sodium or potassium nitrate, string or thin rope made from cotton, linen, jute, hemp, or other vegetable fiber material.

This delay device is usually prepared by boiling the string in a pot of soapy water for about five minutes. The soapy water then is poured away and the pot is refilled with clean water and brought to boiling again. This latter process is repeated three more times, for a period of not less than five minutes each time.

After pouring out the last pot of water, the pot is refilled with about a quart of water, to which is added about 10 spoonsful of sodium nitrate or potassium nitrate.

This mixture is now heated and stirred to dissolve the nitrate, after which the mixture is boiled for another five minutes.

After the five minutes boiling time, the heat is removed and the pieces of string are hung up to dry. To be certain that the string dries thoroughly, at least a three-day (72 hour) drying time should be allowed. Delay time is determined by test-burning measured sections.

An alternative to home-made fuse is model rocket fuse. Typically, it will burn at a rate of between 30 and 40 seconds per foot. As shown in Figure 3-3, it is dark green and comes packaged in a plastic or cellophane wrapper

Figure 3-3.

The burn pattern is shown in Figure 3-4 and will usually leave a trace of "sputtering and spattering."

Delay #4

Candles

Material to be Ignited: Flammable liquids and some ordinary combustibles.

Ignition Source: Open flame.

Materials Needed: Candles.

Candles are usually not utilized in situations where the delay is to be less than one hour. Generally, candle delays are used for time frames between one and eight hours.

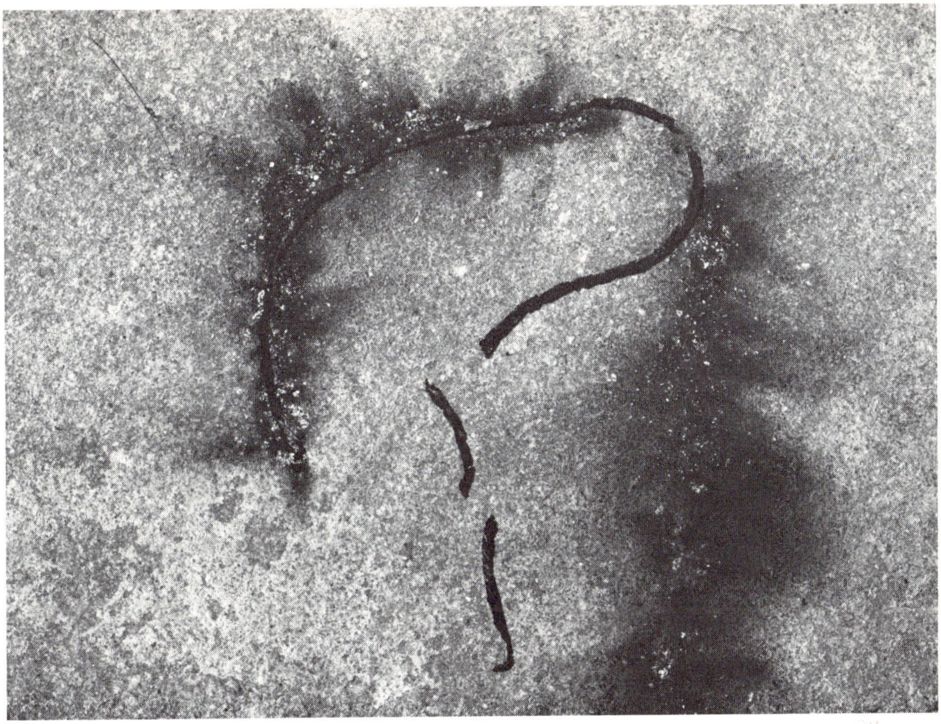

Figure 3-4.

Although this delay works in cold as well as hot weather, it is susceptible to being extinguished by drafts or strong winds. Another disadvantage is that if the candle is not properly affixed at its base, it could float upward in melting wax and not ignite the material it is intended to ignite.

The delay time varies depending on the size and type of candle. To determine what the time may be with any given candle, it is customary to choose two candles of the same size and type and "practice burn" one of the candles which is marked off at one-inch or half-inch intervals timing how long it takes to consume a desired length.

The presence of wax should be an indicator that further investigation is warranted.

4

ELECTRONICS

INTRODUCTION

This chapter deals with both time delay and direct ignition devices. The chapter is divided into three sections: (1) Powerless; (2) D.C. Powered; and (3) A.C. Powered. The premise of all these devices is that an open flame device (e.g., match, electronic igniter) is the ultimate source of ignition. Why go to elaborate electronic means to ignite a fire when a match and gasoline work just as well? The answers are numerous.

Electronic ignition provides ignition sources not otherwise available through 'ordinary' methods. Electronics provides a means to remotely trigger the device, thereby creating the opportunity for the arsonist to establish an alibi. Electronic triggering also provides delays of anywhere from seconds to days. Electronics can also be used to have the 'victim' trigger the fire origin.

Another reason for using electronic devices is that detection is difficult. Few fire marshal offices or police departments have electronics experts available to the Arson Squad or initial scene investigation teams. The most likely components to be found in fire debris where electronics was the ignition source would probably be relay coils, relay frames, and steel battery tops. Smaller components, such as resistors, transistors, and capacitors, will be destroyed if they are in the point of origin, or if the fire was totally destructive. If a remote triggering device is used, there may be wires leading to it and the various components may not be destroyed. As with the components mentioned, circuit boards (usually phenolic) may or may not survive.

Any electric delay/triggering device described in this chapter can be used to activate the other initiating devices in this text.

Direct Powered Photocell/Relay

A photocell is considered "powerless" in that it does not require a power supply for activation. The schematic for the device is shown in Figure 4-1.

The principle behind a photocell (PC on the schematic) is that the device will convert luminous energy into electric energy. The photocell is also known

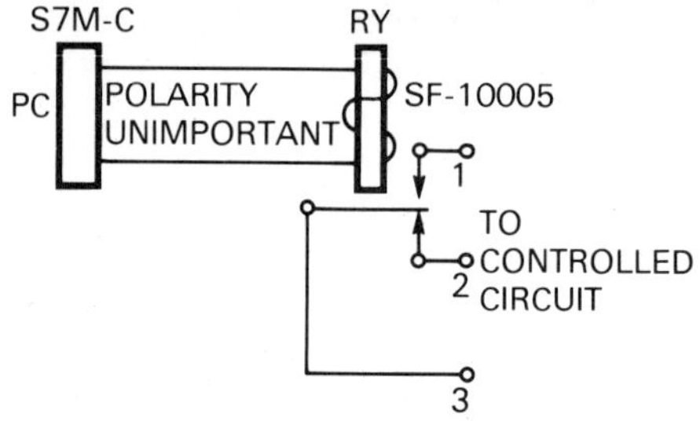

Figure 4-1. Turner: Solar Cells and Photocells. Howard W. Sams, p 38, 1975.

as a light variable resistor in that current flow through the cell is directly dependent on the amount of light striking the face of the cell.

The two basic types of photocells are the photovoltaic cell (which produces d.c. voltages in proportion to the incident light) and the photoconductive cell (also known as a photoresistive cell), which changes internal resistance inversely with the amount of light striking the face of the cell.

Photovoltaic cells normally utilize selenium or silicon as the light sensitive material, whereas most photoconductive cells use cadmium sulfide, lead sulfide, or indium antimonide.

For the circuit shown in Figure 4-1, the output of the photocell directly drives the relay. No battery or other power supply is needed. Depending on the type of photocell used, light from a 75-watt incandescent lamp will provide sufficient illumination, providing the lamp is close enough to the cell. Manufacturer's specifications and testing will provide all the data and distances necessary for operation of the device.

Also note that the relay shown has two sets of contacts. One set is "normally open" (N.O.) while the other set is "normally closed" (N.C.). This sets up the potential for using the circuit to "hold" a circuit shut as needed, with the "opening" of the circuit being necessary for triggering of the device. The reverse is also true.

Any relay used in circuits described in this text would be capable of handling at least a quarter of an amp. If that amperage is not high enough, a relay with that minimum rating could always be used to drive a larger relay.

When using photocells, remember that the sun could be the potential light source. This simply requires proper positioning of the photocell where the sun will strike it.

A porch or crawl space with a lattice work covering is a typical example. Careful observation of shadows at given times will provide data on where to place the cell so that sunlight strikes it at the desired time.

The normally open contacts are connected to the initiating device. The cell is placed so it is in shade until illumination (sunlight) "closes" the normally open contacts. This completes the circuit, causing the device to "fire."

Conversely, setting up the circuit the opposite way requires darkness for activation. Darkness, in this method, unlatches the relay and returns it to its normally closed position which activates the device. Figure 4-3 is a schematic of this set-up.

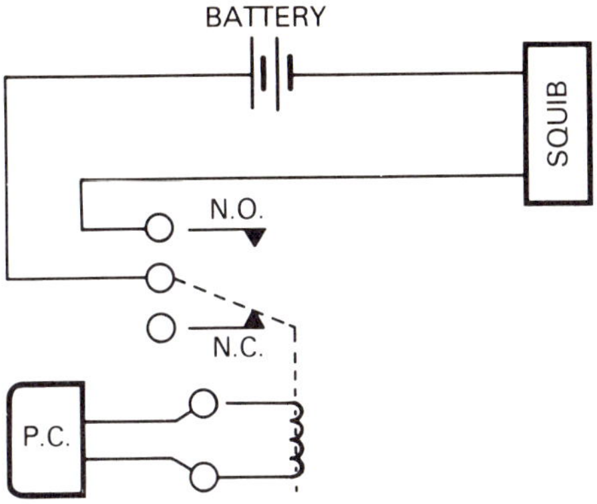

Figure 4-2.

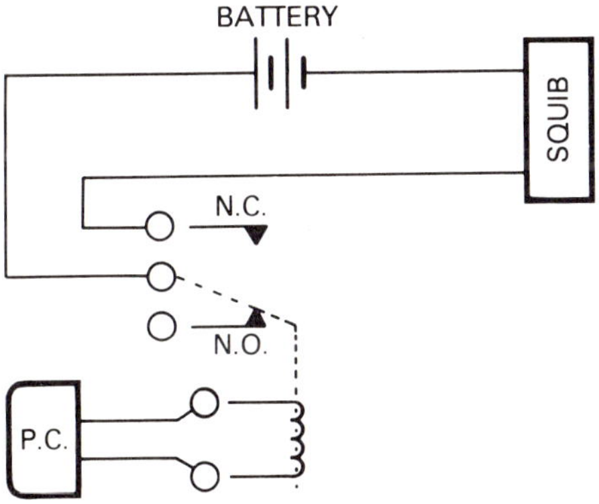

Figure 4-3.

Transistorized Photocell Circuit #1

This circuit, using a transistor to amplify the photocell output, is designed to drive a low-current relay that could directly trigger a device or drive another relay. The basic circuit is shown in Figure 4-4.

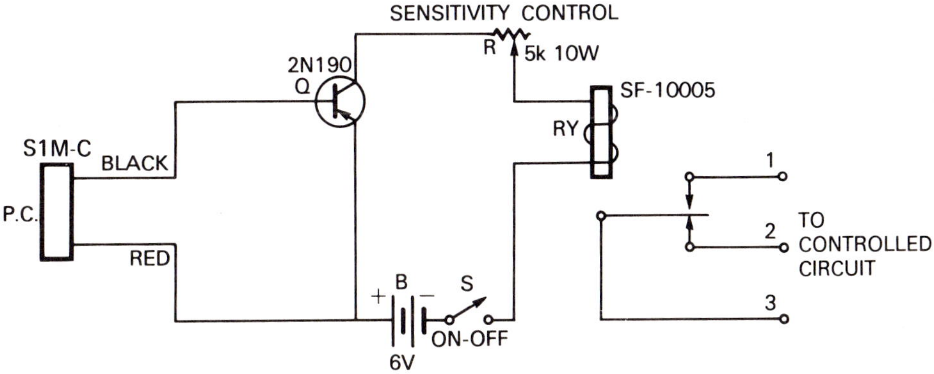

Figure 4-4. Turner: Solar Cells and Photocells. Howard W. Sams, p 40, 1975.

In this instance, the photocell is an International S1MC, while the transistor (Q1) is a Radio Shack® 276-2007 (PNP) and the relay is a Sigma® 5F1000S. As in the previous circuit, use of normally open relay contacts will provide triggering in one direction, while use of the normally closed contacts can provide a completed circuit in the opposite direction.

In theory, 30 foot-candles of light should be sufficient to activate the circuit. As a point of reference, a 60-watt tungsten-filament lamp will deliver approximately 50 foot-candles at a distance of one foot from the lamp. Other light values are given by the various lamp manufacturers in their descriptive literature.

Transistorized Photocell Circuit #2

For this circuit (Figure 4-5) the output of a small silicon photocell is amplified by means of the transistorized d.c. amplifier to drive a higher current relay than the cell itself could operate.

The relay in this circuit could be a relay in an air conditioner, refrigerator, or furnace (see Apppliances Chapter) which will "spark" in an explosive vapor atmosphere, creating the impression of an electro-mechanical failure.

Again, the photocell is an International S1MC with a Radio Shack® PNP Transistor (276-2007); however, now a nine-volt battery is used instead of a six-volt.

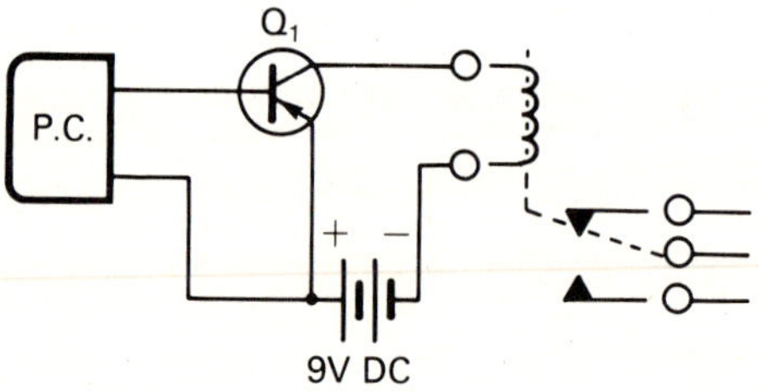

Figure 4-5.

Flashlight and Photocell

This device utilizes four "D" cells, one nine-volt battery, an SPDT subminiature PC Relay, and a photocell. The flashlight houses all of the components. Since the flashlight is plastic, the odds are it will be destroyed or at least melted to the point where it would simply look like a flashlight. The leads most probably would burn off in a successful fire.

The schematic is:

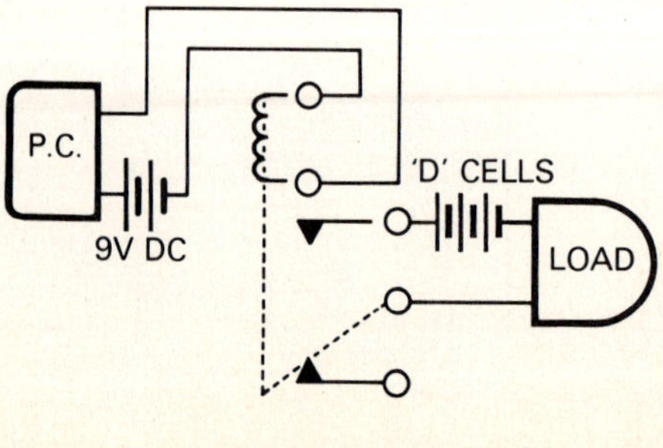

Figure 4-6.

A photograph of the device is shown with an electric squib and a solar igniter, either of which could be used with a device of this nature.

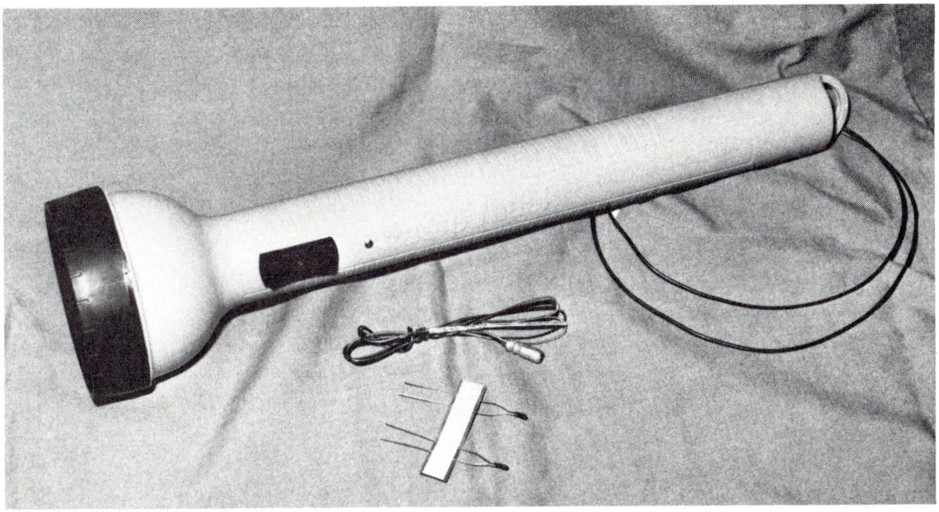

Figure 4-7. Completed Package.

The next picture (Figure 4-8) shows the disassembled unit. Look for two different types of batteries ("D" Cell and nine-volt) in the same housing.

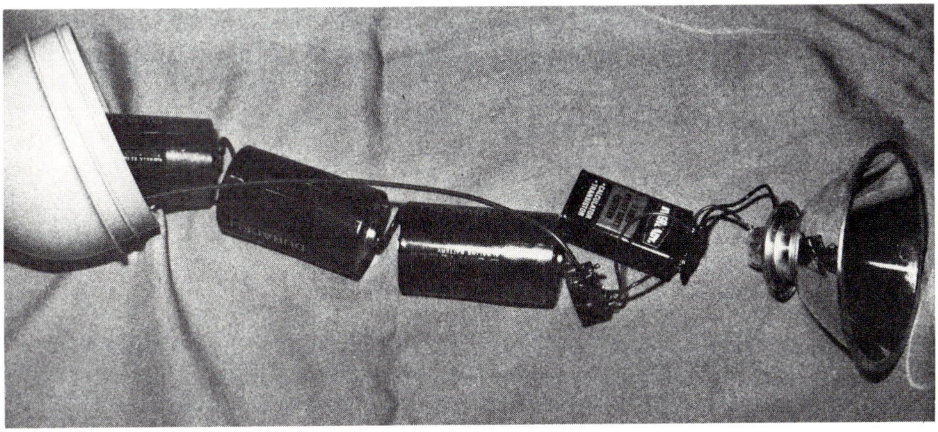

Figure 4-8. Disassembled Unit.

One additional clue is the absence of a bulb in the reflector section of the flashlight. It was replaced with a photocell

The most logical set-up of this device is to lay it where sufficient sunlight (or artificial light) will strike the photocell and activate the circuit.

In this circuit the resistance of the photocell is lowered to the point where current flows through the photocell from the nine-volt transistor battery. This subsequently allows current flow through the coil portion of the relay, which, in turn, "pulls" the contact points from the normally closed side to the normally

Figure 4-9. Photocell in lieu of bulb.

open side. This now allows current to flow from the "D" cells through the igniter circuit. The relay is a Radio Shack® 275-216, which requires only five volts d.c. to operate. It has an internal resistance of 56 ohms, a current requirement of 90 ma, and a contact rating of 1 amp at 125 volts a.c.

Light-Controlled SCR Triggering

Since a mechanical relay possibly will unlatch, an alternative latching mechanism is the silicon-controlled rectifier (SCR). The principle of the SCR is that, in its "normal" state (untriggered), it will block a voltage applied in either direction. Once "enabled" by a voltage or signal applied to the gate electrode, the SCR will immediately "latch" or conduct, even if (or after) the control signal or control voltage is removed, reduced, or reversed. The only way to stop conduction is to remove the anode supply voltage.

The schematic symbol for an SCR is:

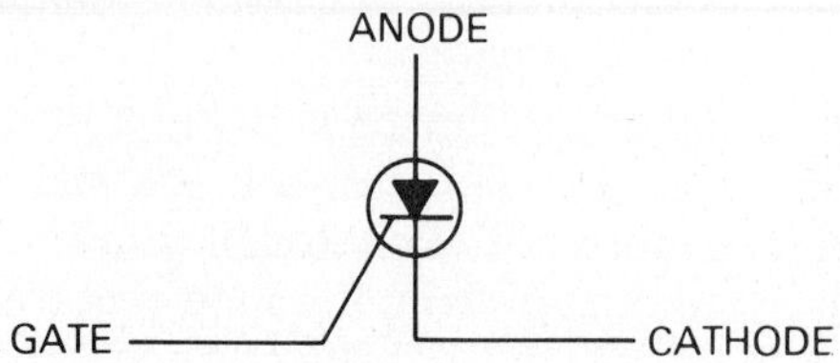

Figure 4-10.

With this principle in mind, it becomes a simple matter to design a circuit using very few components, which could be triggered by a photocell, timer, touch switch, or other device. The following circuit could be built on a 1 inch × 2 inch (max) circuit board.

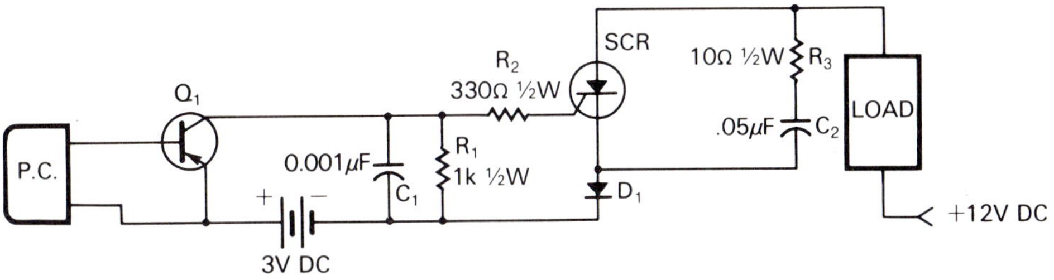

Figure 4-11.

The SCR turns on when the input goes positive with respect to the cathode. R1 eliminates falsing from line noises. R2 is a current limiting resistor. Diode D1 helps increase noise immunity. R3/C2 is a clamping circuit to prevent the SCR from "firing" when power is applied. The SCR is a 2N3228 or equivalent. Q1 is any PNP transistor.

The "load" could be nothing more than several "barn burner" matches or some solar igniters (available from hobby stores, used to ignite model rockets).

An alternative circuit is shown in Figure 4-12.

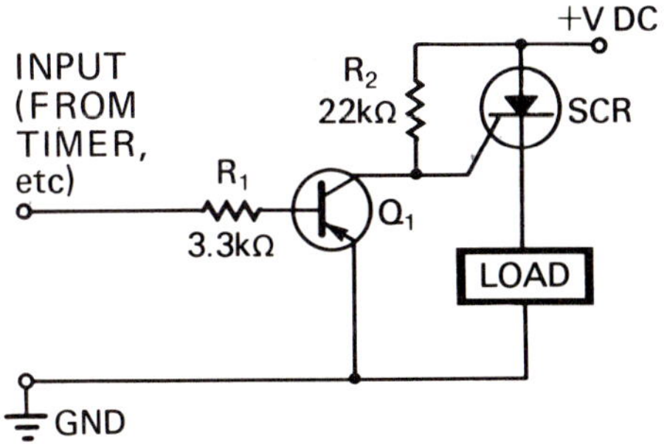

Figure 4-12.

Timing/Delay Circuit

This circuit provides a delay for triggering a circuit. In this instance a 555-timer chip keeps the voltage at zero (on the output) until a certain time interval has passed. Switch (SW-1) must be opened to activate the circuit. Figure 4-13 is the schematic.

R1/C2 are the timing network components, which can be chosen by experimenting. Delay can be from seconds to hours.

A drawback with this circuit is that the output of the time chip can handle only between 100 and 120 ma, so this timer could not be used (in this configuration) with any load demanding heavier current. To use a higher current device, the following PNP buffer circuit needs to be added.

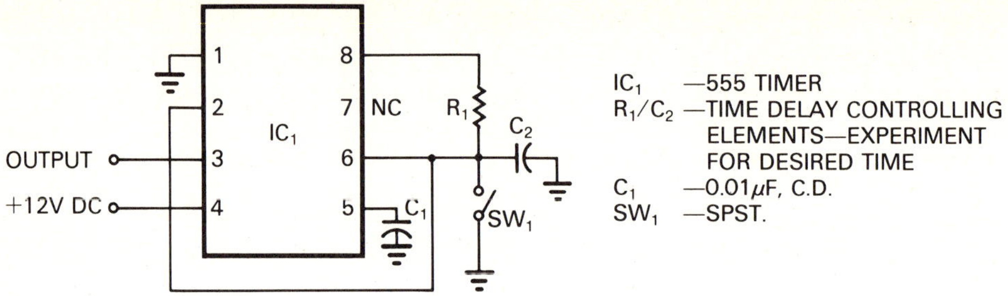

Figure 4-13.

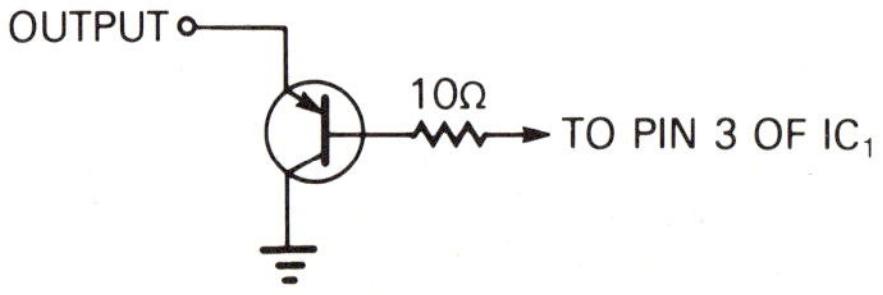

Figure 4-14.

An ECG 127 or WEP 235 transistor can handle up to 10 amps, an ECG 281 transistor up to 12 amps, and a WEP 776 transistor can handle 16 amps.

For reference purposes, the following timing circuits are provided. Circuit boards found at the scene should be "traced" to discover possible timer circuits.

Delay turn-off circuits can be used with normally open relays. When the relay closes, the circuit activates. In this manner a motor or other circuit can continue to function as scheduled. Shut-off provides the ignition. For this reason, delay turn-off circuits will be included in the timer circuit diagrams.

It must also be remembered that, although a particular circuit in and of itself will not be able to "drive" the igniter, a buffer or amplifier stage could be inserted between the timing circuit and the igniter. This amplifier or buffer circuit will increase the current potential to the level needed for the igniter (See Figures 4-15 through 4-28).

NAT/VAT

The noise-activated trigger (NAT) or voice-activated trigger (VAT) represents a somewhat unique device that can be set up to trigger on an incoming phone call, ambient room noise, or various other noise sources. This particular device was given to the author for experimental purposes.

Once again, the SCR is employed as the main triggering control mechanism. The basic circuit is shown in Figure 4-29.

A photograph of the device is presented in Figure 4-30. Note relative size of device as indicated by the ruler.

Testing of the circuit, prior to final connections, is done with a #48 or #49 light bulb (lamp). Resister R1 is adjusted until the light bulb turns on. At that point the resistor is adjusted slightly backward from the previous direction of

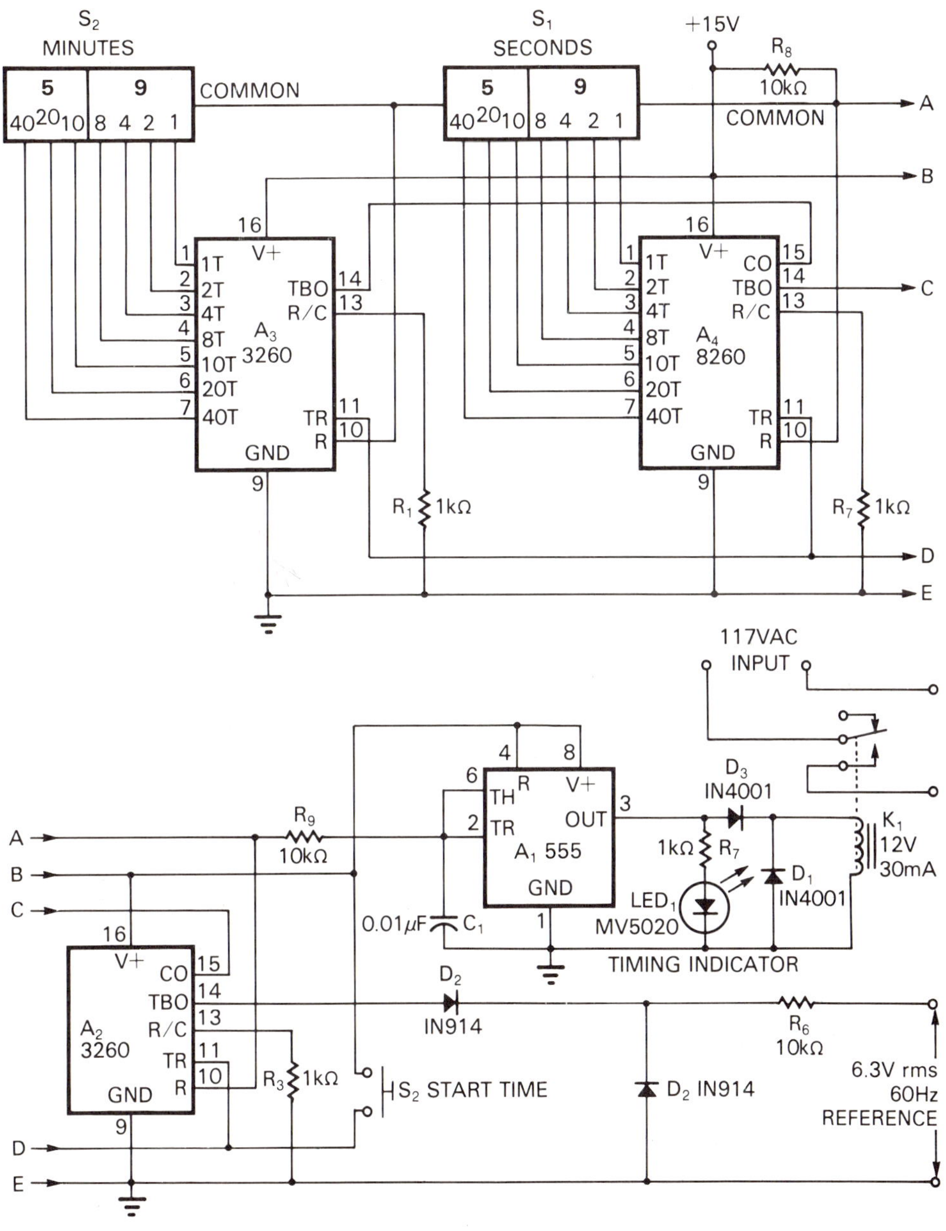

Figure 4-15. APPLIANCE TIMER—Controls intervals up to 1 h in 1-s increments as programmed by thumb-wheel switches S_1 and S_7. Circuit is basically two-stage programmed counter driven by 1-s clock derived from 60-Hz power line. A_1 is connected as divide-by-60 counter triggered by 60-Hz signal developed across D_2. 1-Hz output from A_1 triggers A_2 which in turn triggers A_3, all programmable timers. When S_3 is closed, R_4 bus goes low to start timing cycle. Relay driver A_4 holds relay K_1 closed for application of AC power to device being controlled and energizing of LED to indicate active timing cycle. Applications include uses as kitchen and darkroom timers.—Jung WG: IC Timer Cookbook. Howard W. Sams, Indianapolis, IN, p 214-218, 1977.

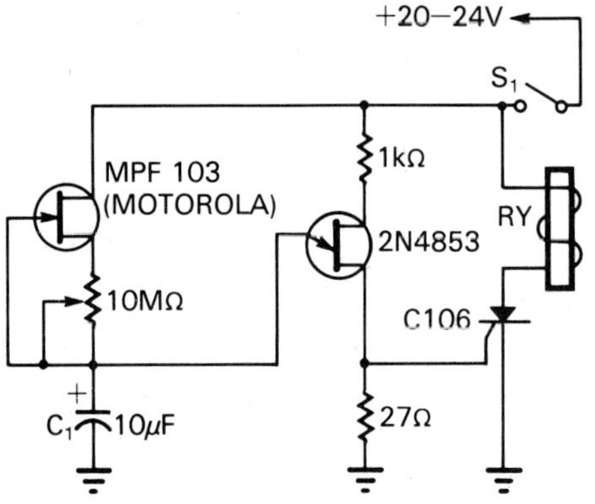

Figure 4-16. 10-min DELAY—MPF103 JFET makes charging current of C_1 constant regardless of degree of charge, to give longer charging time and longer time delay. After delay determined by setting of 10-megohm pot, UJT conducts and discharges C_1 through 27-ohm resistor, triggering SCR and energizing relay. Open S_1 to reset circuit.—I. Math, Math's Notes, *CQ,* April 1974, p 64-65 and 91-92.

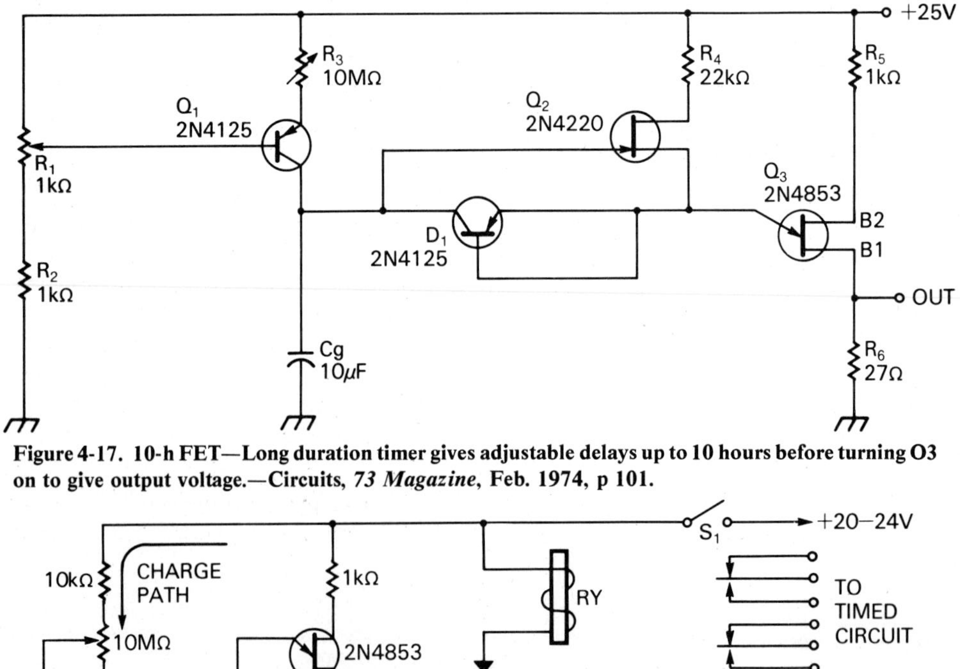

Figure 4-17. 10-h FET—Long duration timer gives adjustable delays up to 10 hours before turning O3 on to give output voltage.—Circuits, *73 Magazine,* Feb. 1974, p 101.

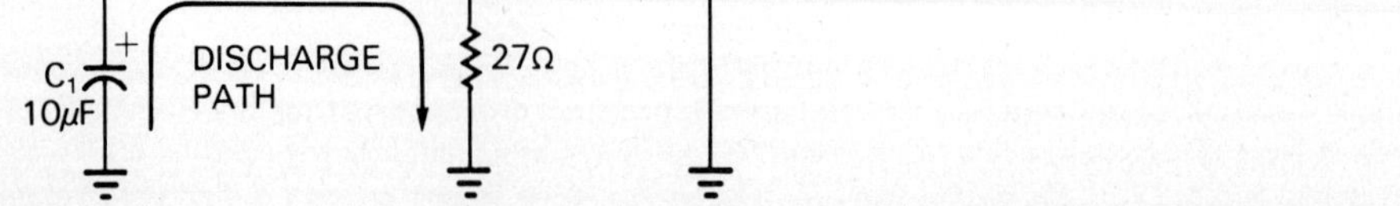

Figure 4-18. 1.5-min DELAY—When S_1 is closed, C_1 begins charging. After delay determined by setting of 10-megohm pot, UJT conducts and makes C_1 discharge through 27-ohm resistor, triggering SCR C106 and energizing relay. Open S_1 to reset circuit.—I. Math, Math's Notes, *CQ,* April 1974, p 64-65 and 91-92.

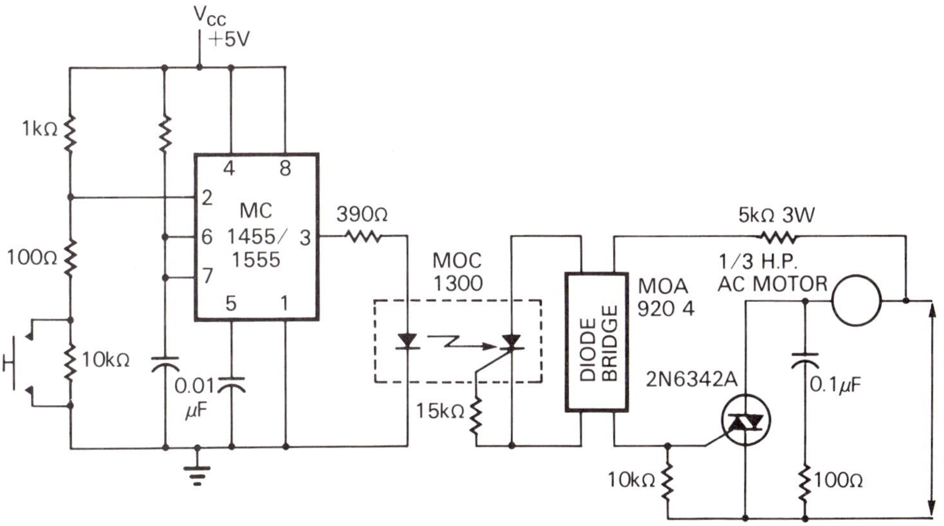

Figure 4-19. LONG TURNOFF DELAY—Combination of timer, optolsolator, and bridge-triggered triac keeps AC motor or other device energized for up to 1 h after control switch is depressed momentarily. Closing of switch drops voltage at pin 2 of timer below ⅓ V_{cc}, making timer output go high and thus turn LED on. At same time, capacitor at pin 7 begins charging. Output remains high until capacitor reaches ⅔ V_{cc}, when output is reset to low state and motor thereby turned off.—T. Mazur, Solid-State Relays Offer New Solutions to Many Old Problems, *EDN Magazine,* Nov. 20, 1973. p 26-32.

turning. This process establishes the "threshold" firing point of the SCR. Power is then disconnected and the unit is ready for service. A telephone, shouting, or other equivalent noise within 10 to 15 feet of the microphone is usually sufficient to activate the unit.

Lines 1 and 2 (incoming) of the telephone lines can also be used as a "direct-wired-signal source" in lieu of the microphone. To do this, the microphone lines are disconnected and the following input circuit would be used (Figure 4-31).

The "load," in either circuit, can be a kitchen match, solar igniter, or any other open flame device the unit is capable of "driving" (Figure 4-32).

In a "total destruction" fire, the most that would probably be available to the investigator is the steel battery top, as shown in Figure 4-33. This alone is not sufficient to be considered as the source of ignition.

Touch Triggering

Utilization of this device to trigger a fire (or explosion) means the intended 'victim' could activate the device accidentally. The principle involved is that a low-resistance device (like a finger) completes an electrical circuit when two adjacent metal strips are placed in contact with one another.

This particular circuit uses a relay with a coil having greater than 250 ohms resistance. It can also be used with an SCR. The circuit is depicted in Figure 4-34.

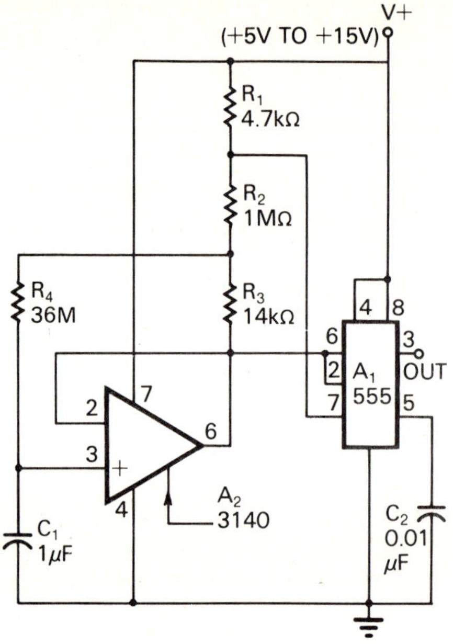

$$T = \frac{R_2 + R_3}{R_3} (1.386\ R_1C_1)$$

$$T = 100R_1C_1$$

(AS SHOWN, T = 1 HR.)

Figure 4-20. MICROSECONDS TO HOURS—Timing range of 555 is increased 100 times by using 3140 FET-input opamp in circuit that effectively multiplies values of timing components and buffers timing network. Pin 7 of 555 switches applied voltage of timing network between V+ and ground. Ratio selected fro R_2-R_3 can be varied over wide range to change multiplying ratio, provided square-wave voltage across R_3 is at least 50 m V. Output is essentially pure square wave. Supply voltage is not critical.—W. G. Jung, Take a Fresh Look at New IC Timer Applications, *EDN Magazine,* March 20, 1977, p 127-135.

An alternative circuit possibility uses a 555-timer chip and is schematically shown in Figure 4-35.

Since numerous items presently being marketed are controlled by touch switching, the following circuits are provided as reference materials (Figures 4-36, 4-37, 4-38, 4-39 and 4-40).

Batteries, Telephone, and Matches

This is one of the simpler devices to prepare. However, it does have one drawback in that delay of ignition is virtually impossible to control. The circuit consists of one new "D" cell battery, wire from a telephone ringer coil, and a "barn burner" match or other open flame device such as a solar igniter (see next section).

A "D" cell battery is placed in a battery holder of the appropriate size. (An alternative would be to drive two nails in a wooden framework and wedge the

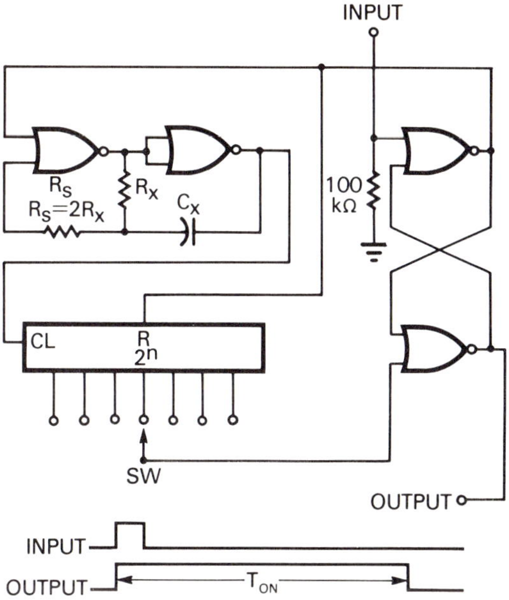

Figure 4-21. MICROSECONDS TO HOURS—Simple CMOS circuit serves as time-delay switch and general-purpose timer. Gated oscillator and latch are obtained from CD4001 quad two-input NOR gate, and 14-stage counter uses CD4020. T_{ON} is function of oscillator frequency as determined by R_XC_X and proper 2^N output from counter. Pulse applied to latch input enables oscillator and counter. Latch output remains high until counter resets latch at end of count selected by switch. Further decoding is required for count or time variations finer than power of 2.—J. Chin, Low-Power Counter Is Programmable over Wide Range, *EDN Magazine,* March 20, 1974, p 83.

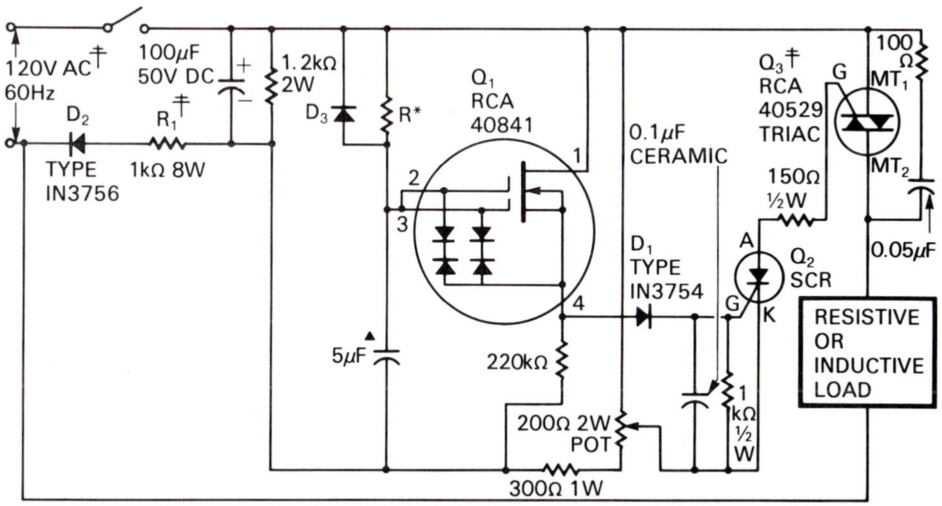

Figure 4-22. 0-5 min DELAY—Value of resistor R controls duration of time delay provided by RCA 40841 dual-gate FET in SCR trigger circuit, with SCR in turn serving to trigger triac for handling high-current resistive or inductive AC loads. Maximum delay of 5 min is obtained when R is 60 megohms (IRC type CGH or equivalent resistor). Timing is accurate within 10% over temperature range of −25°C to +60°C. D_1 should be rated 60 V. Use any SCR capable of handling triac trigger current, rated 60 V.—"Linear Integrated Circuits and MOS/FET's," RCA Solid State Division, Somerville, NJ, 1977, p 435-437.

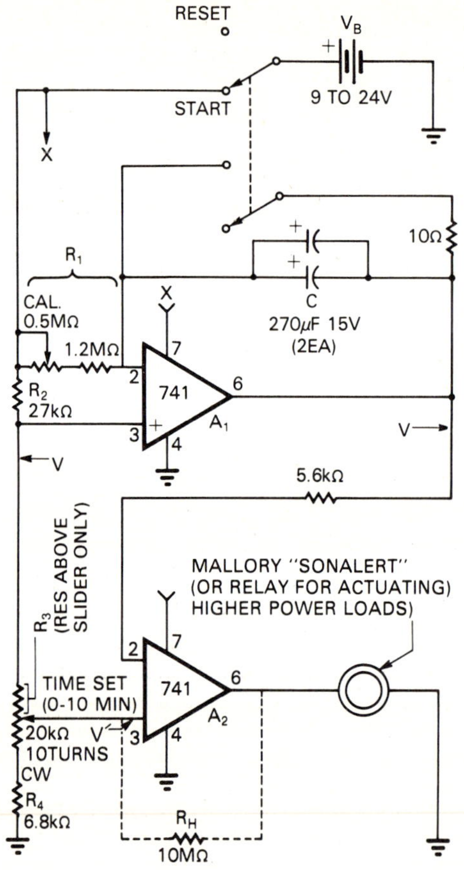

Figure 4-23. 0-10 min WITH 1-s ACCURACY—After calibration, accuracy is independent of battery voltage because source voltage affects charging voltage of C and threshold of comparator A_2 equally. Time delay t for timer is CR_1R_3/R_2.—M. Strange, Simple Electronic Timer is Compact and Accurate, *EDN Magazine*, April 20, 1973, p 89 and 91.

battery between the nails. Following this, the ringer coil is removed from the telephone and a few inches of coil wire is unwound (see Figure 4-41).

The match is "prepared" (if a match will be used) by drilling a small hole through the head of the match and wrapping the ringer coil wire through and around the head of the match, as shown in Figure 4-42.

Prior to connecting the wires to the battery assembly, the area is "presoaked" with a liquid flammable, usually in an atomizing fashion so as to have vapor throughout the area. Final assembly is then made between the wire ends and the battery. Ignition usually occurs in 30 to 60 seconds unless a delay device is placed in one of the lead lines between the match and the battery.

The obvious place to set up this device is in any appliance which uses batteries, e.g., portable television, radio, tape recorder, or other similar device.

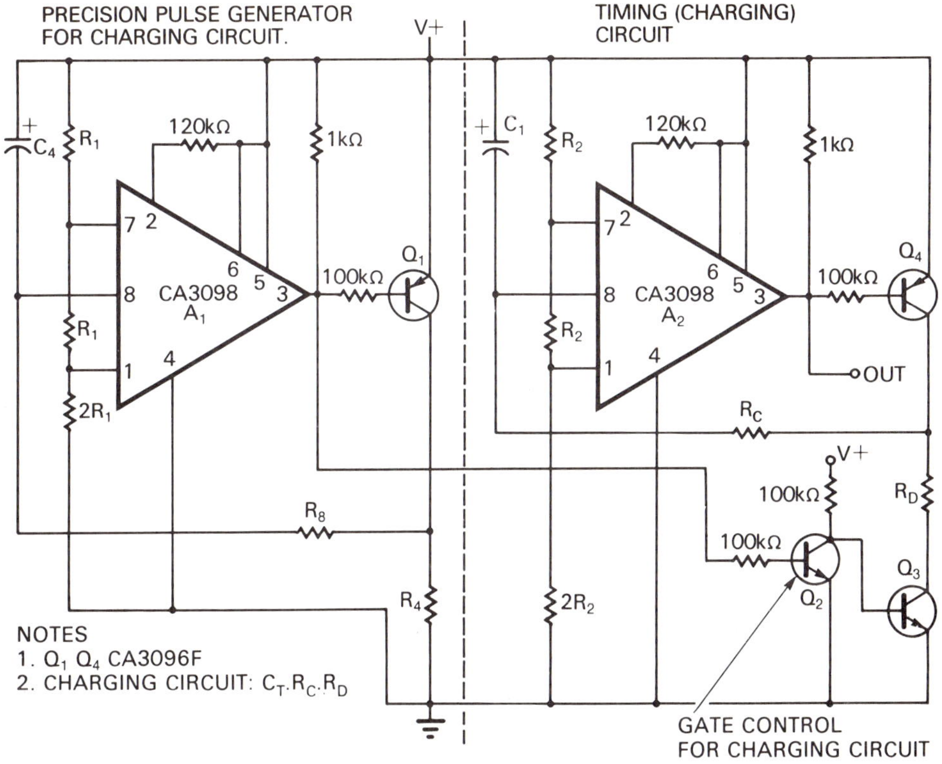

Figure 4-24. LONG INTERVALS WITH SMALL C—Use of two CA309C dual-input precision level detectors eliminates need for expensive high-capacitance low-leakage timing capacitors when delay intervals of several hours are required. For 4-h timer, C_1 is only 16 *m*F if R_C is 22 megohms and R_0 is 100 kilohms. Article traces circuit operation and gives design equations.—G. J. Granieri, Precision Level Detector IC Simplifies Control Circuit Design, *EDN Magazine,* Oct. 5, 1975, p 69-72.

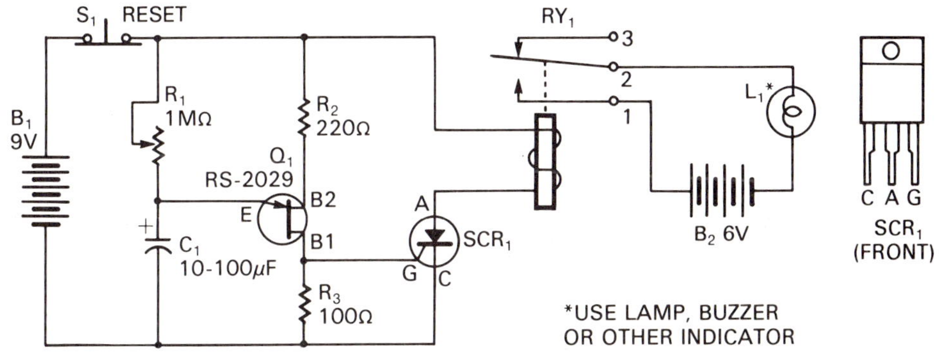

Figure 4-25. UJT-SCR TIMER—Use of large capacitance for C1 in simple UJT relaxation oscillator provides time delay action for triggering SCR controlling relay. R1 provides convenient adjustment of delay. SCR can be 6-A 50-V Radio Shack 276-1089. Relay is 275-004.—F. M. Mims, "Semiconductor Projects, Vol. 2," Radio Shack, Fort Worth, TX, 1976, p 50-61.

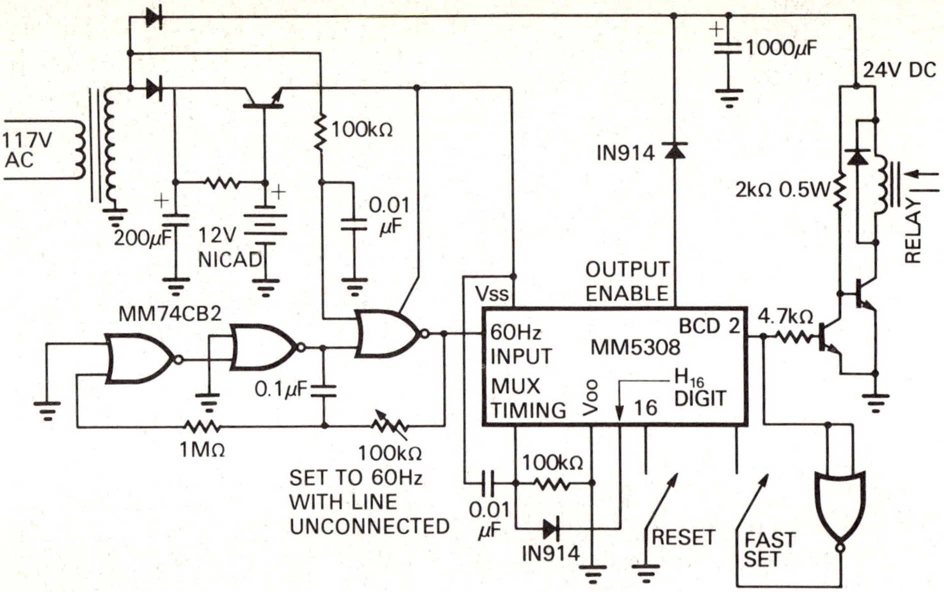

Figure 4-26. FAIL-SAFE LIGHT TIMER—National MM5309 clock IC is used as timer in circuit that maintains timing with adequate accuracy during periods of power-line failure and returns automatically to 60-Hz line as soon as power is restored. Applications include control of lights in unoccupied home. Timing action turns on lights for 4-h period every 24 h. When power is applied, internal multiplex circuit strobes each digit until digit with connected diode is accessed. This digit stops multiplex scanning, and BCD outputs present data from selected digit as control waveform whose edges determine timer data.—"MOS/LSI Databook," National Semiconductor, Santa Clara, CA, 1977, p 1-74-1-77.

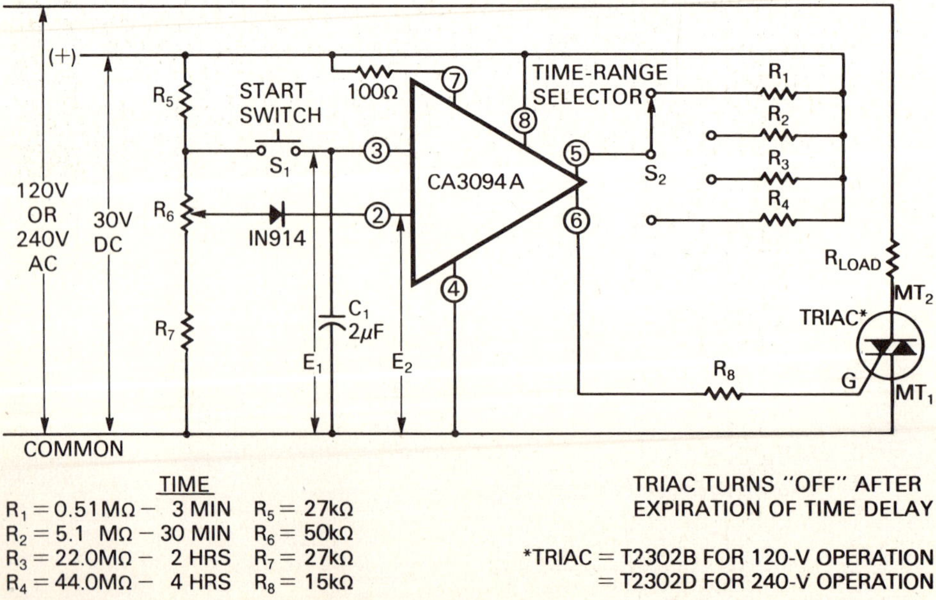

Figure 4-27. 3 min TO 4 h—Presettable analog timer achieves long time intervals by discharging C_1 into input terminal 3 of CA3094 programmable opamp, which provides sufficient output current for driving thyristors and other control devices.—"Circuit Ideas for RCA Linear ICs," RCA Solid State Division, Somerville, NJ, 1977, p 8.

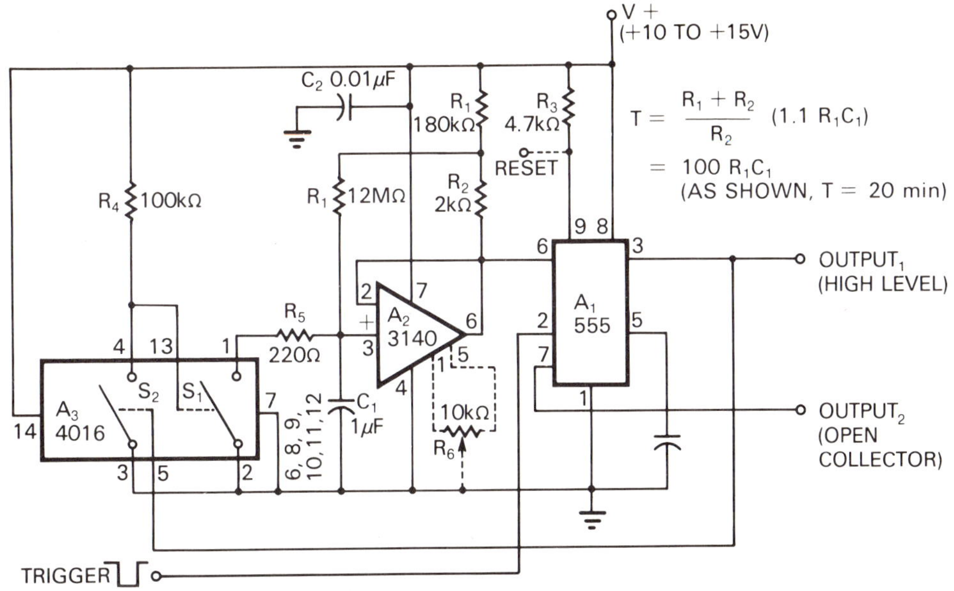

Figure 4-28. LONG-DELAY 555 MONO—FET-input 3140 opamp is used to multiply effective values of timing components R_1 and C_1, eliminating need for high-value precision resistor and large low-leakage capacitor. Combination performs as standard 555 mono except that timing equation is $T = 100R_1C_1$ (for condition wherein division resistors R_1 and R_2 are chosen for 91-to-1 operation). Circuit has uncommitted open-collector output from pin 7 of 555, which can be referred to any voltage from 0 to +15 V. Pin 4 is between pins 6 and 8 on 555.—W. G. Jung, Take a Fresh Look at New IC Timer Applications, *EDN Magazine*, March 20, 1977, p 127-135.

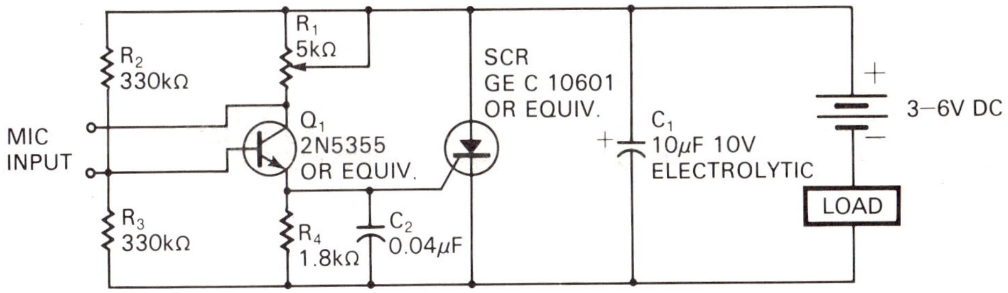

Figure 4-29.

Placing the "devices" on a shelf, with appropriate "presoak" of the device and the shelving area, should lead to a reasonably successful fire. Electrical problems with the appliance can then be falsely given as the cause of the fire. The schematic of this circuit is presented in Figure 4-43.

Model Rocket Igniters

These devices are used by model rocket enthusiasts and are readily available from most hobby shops. The most success has been achieved with the "Solar Igniter®," pictured in Figure 4-44.

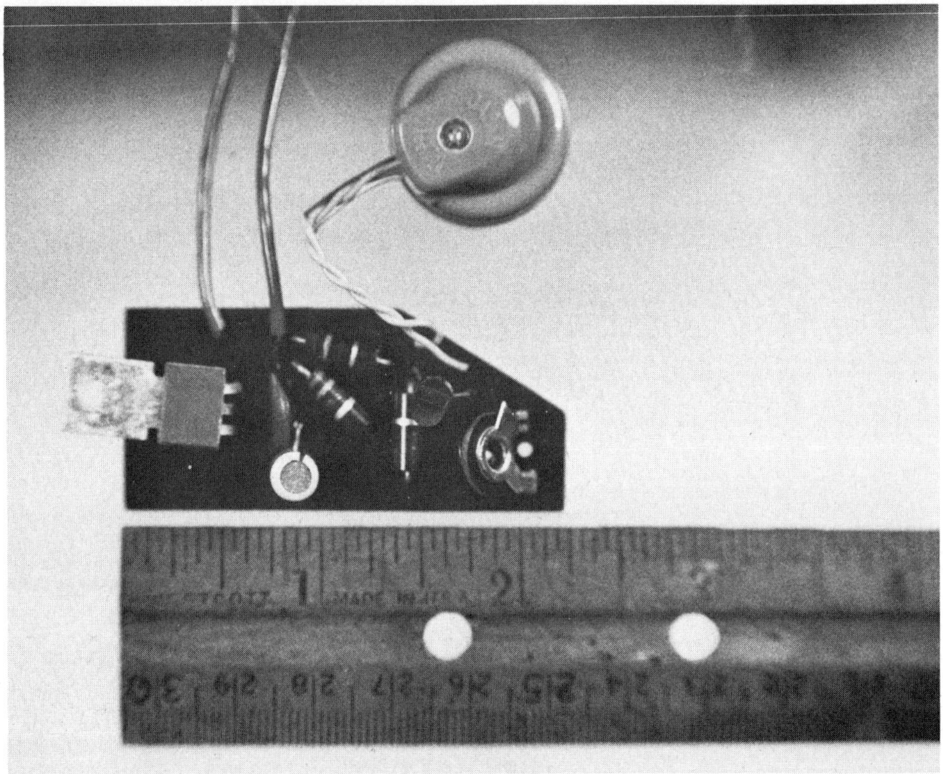

Figure 4-30.

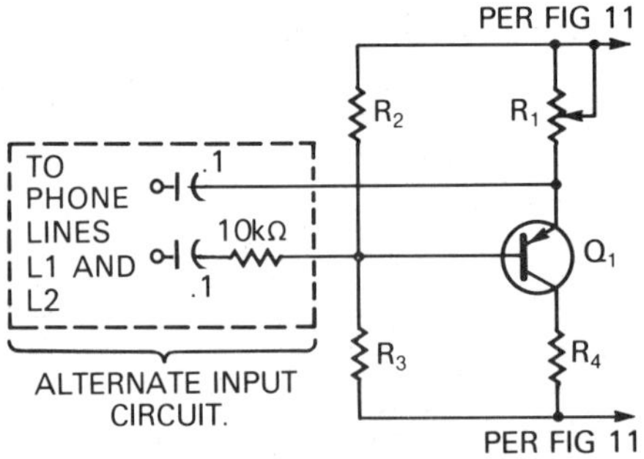

Figure 4-31.

Inside this igniter is nichrome wire, similar to that used in toasters, etc. As shown in Figure 4-44, the igniter resembles a match.

After ignition and burn-off of the exterior covering material, the nichrome wire continues to "glow," quite hot, until current is removed or until the nichrome wire melts in half.

Note also in Figure 4-44 that the igniter comes "supported" with two pieces of paper glued together. This not only acts as support but also keeps the

Figure 4-32.

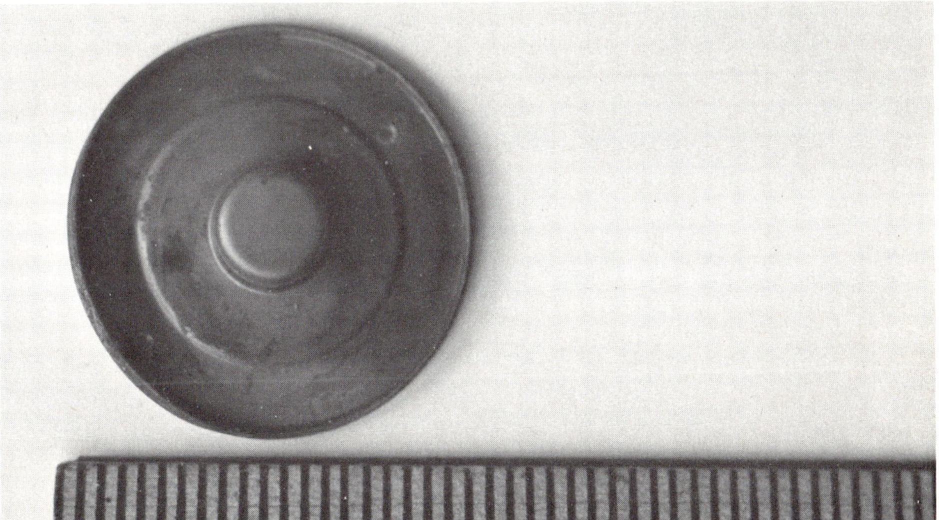

Figure 4-33.

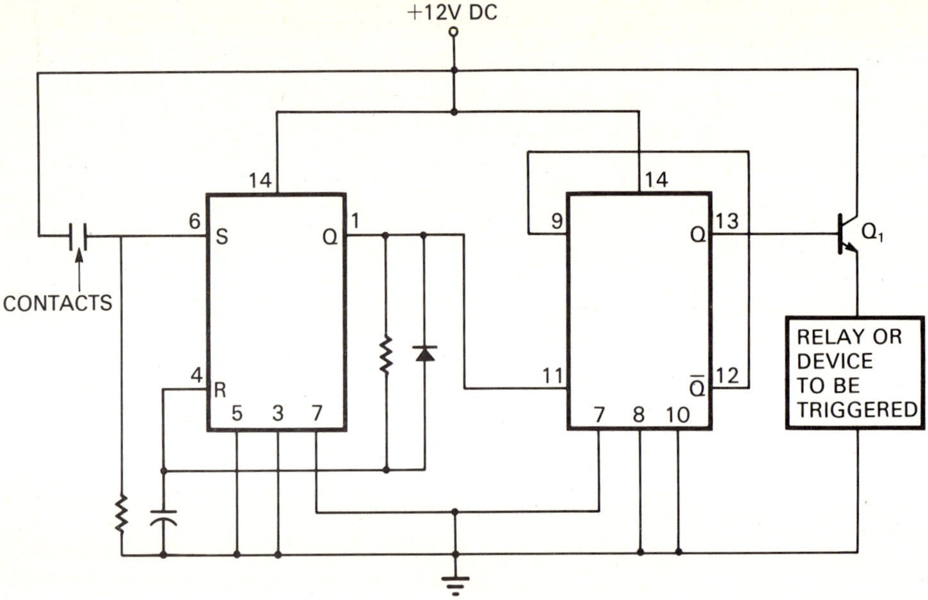

Figure 4-34.

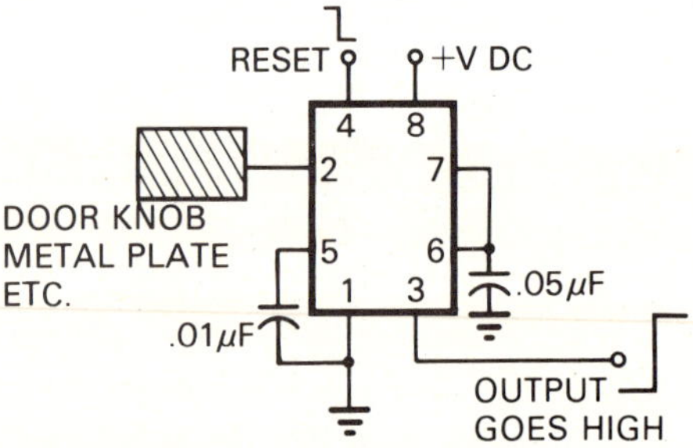

Figure 4-35.

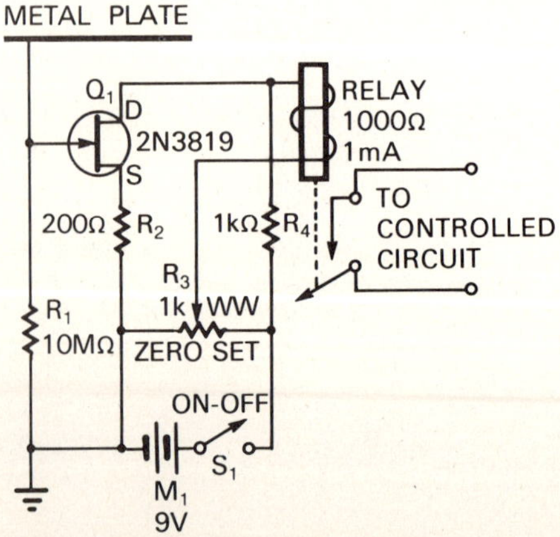

Figure 4-36. TOUCH-PLATE RELAY—When 2-inch diameter disk of sheet metal or foil is firmly touched with finger, stray noise picked up by body and coupled into 10-megohm gate circuit of FET is sufficient to boost drain current to about 1.7 mA and close relay. Delayed dropout can be obtained by placing capacitor in parallel with relay coil; delay is about 0.8 a per 1000 μF of parallel capacitance.—R. P. Turner, "FET Circuits," Howard W. Sams, Indianapolis, IN, 1977, 2nd Ed., p 104-105.

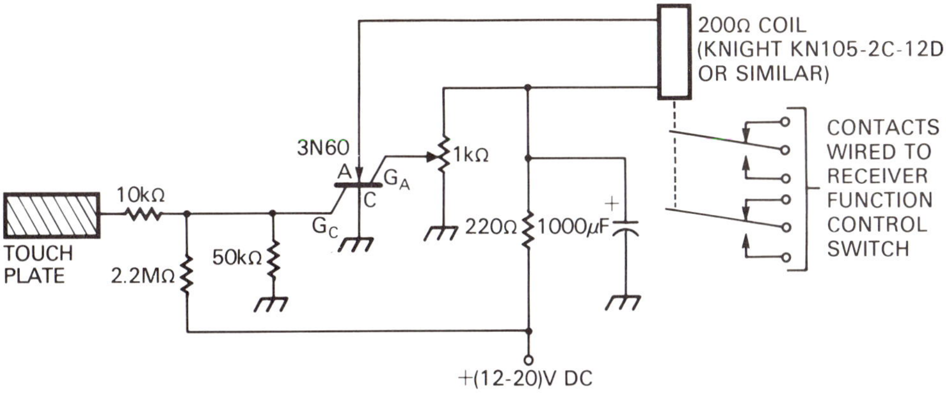

Figure 4-37. TOUCH SWITCH—Performs function of switch by means of relay contacts when SCR is triggered by placing finger on touch plate. Values shown keep relay energized for 5-10 s after touch. Developed as replacement for switch-type controls on amateur radio receiver. Once SCR has fired, it conducts until charge on 1000-μF capacitor decreases enough to drop SCR current below minimum for conduction.—J. J. Schultz, Rapid Receiver Control Switching, *73 Magazine*, Dec. 1973, p 67-69.

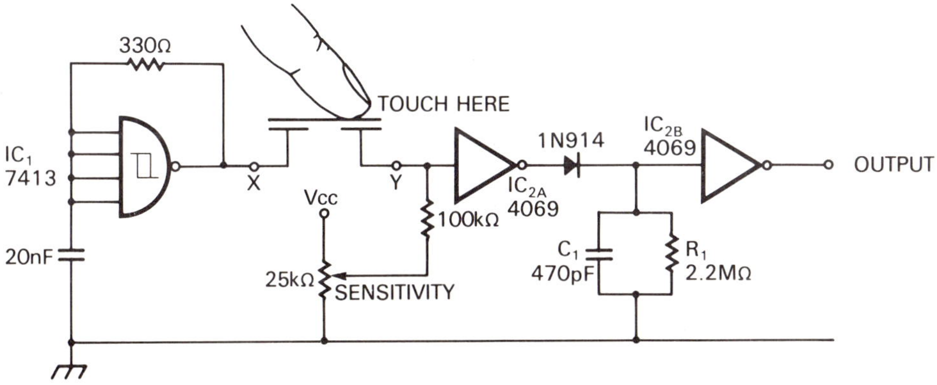

Figure 4-38. LOGIC SWITCH—Can be used with battery-powered circuits because CMOS touch switch does not require body pickup of AC line hum for switching action. Schmitt trigger IC_1 forms 100-kHz oscillator. IC_{2a} amplifies oscillator output and charges C_1 through diode. When sensor is touched, oscillator output is severely attenuated, making c_1 discharge and thereby changing output state of level detector IC_{2b}. Sensor is 1-inch-square of double-sided printed-circuit board with lower side divided into two equal sections.—N. Sunderland, C.M.O.S. Touch Switch, *Wireless World*, May 1978, p. 69.

main leads separated from each other so as to preclude a short circuit between the leads.

Ignition of the solar igniter can be from a six-volt (d.c.) combination of four AA cells, two "C" cells, or one "D" cell alkaline battery. Six- or twelve-volt lantern or car batteries will also work. Likewise, any six- or twelve-volt d.c. source capable of producing about two amperes will work. These sources could be found inside certain radios, televisions, stereos, and other similar appliances.

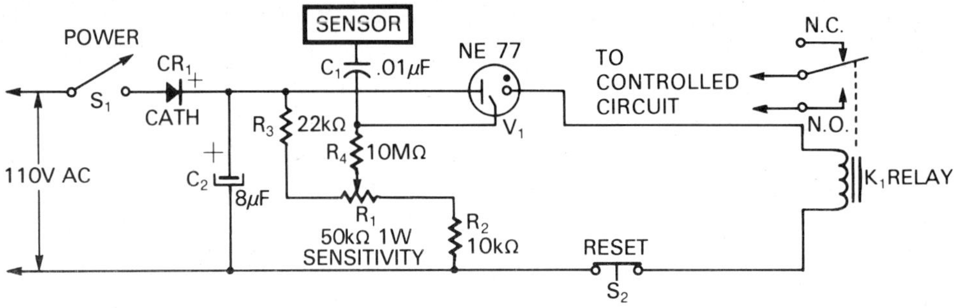

Figure 4-39. TOUCH SWITCH—Uses NE-77 neon lamp, which is similar to NE-2 but has third electrode for triggering. When person touches metal sensor plate of switch, AC voltage picked up by body is applied to trigger electrode of neon, making it fire and energize 5000-ohm relay K1 (Potter & Brumfield RS5D or equivalent). Relay remains energized until S2 is opened to reset circuit. Adjust R1 so voltage applied to center electrode of V1 is just below trigger point.—J. P. Shields, "How to Build Proximity Detectors & Metal Locators," Howard W. Sams, Indianapolis, IN, 2nd Ed., 1972, p 52-55.

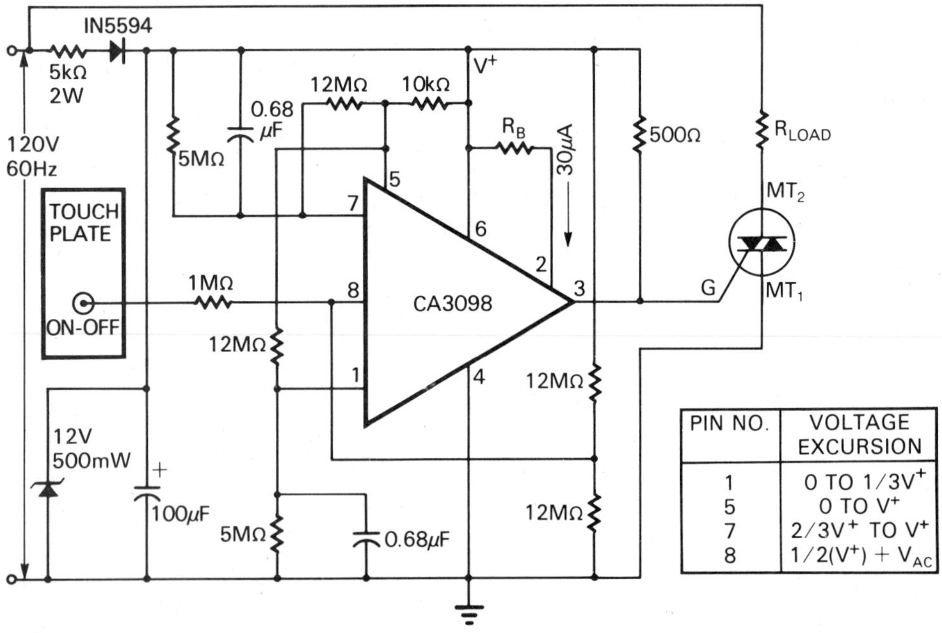

Figure 4-40. TOUCH SWITCH—Small AC signal momentarily introduced by finger contact on touch plate causes voltage at pin 8 of CA3098 dual-input precision level detector to be greater than high reference voltage. This toggles memory flip-flop in IC, making voltage high at pin 5. Voltage at pin 7 then increases exponentially to V+ in about 10 s. This 10-s delay is maximum that button can be touched; longer touch makes system oscillate between ON and OFF states until finger is removed. Shorter touch energizes load, placing pin 7 at V+. Next touch of plate turns circuit off.—G. J. Granieri, Precision Level Detector IC Simplifies Control Circuit Design, *EDN Magazine,* Oct. 5, 1975, p69-72.

Analysis of fire debris probably would not reveal much direct evidence, with the possible exception of a piece of the nichrome wire, which would be extremely difficult to locate.

Figure 4-41.

A.C.-Powered Photocell/Relay

The necessary items in this device are a photocell, (such as Clairex® CL5M4L), an a.c.-powered relay (Magnecraft® W588ACPX-16) and 120 volts a.c. at 60 ma. The schematic is given in Figure 4-45.

In this example, the relay contacts will handle between five and ten amperes, so many ignition devices can be operated directly off of the relay contacts.

To obtain a delay of several days, a stepping relay could be driven by the main relay. Each time light activates the circuit, the stepping relay moves one more step along the chain, until the appropriate pair of contacts is made, at which time ignition would occur through the completed circuit.

The relay could also be the control relay for an air conditioner or other appliance. In this case, the area around the relay would be presoaked. Vapors would ignite from resulting sparks of "making" and "breaking" of the contacts. (This presumes that any protective covers over the relay/contacts are removed to expose the contacts to the atmosphere.)

Look for the photocell when a fire is attributed to mechanical breakdown or short circuit.

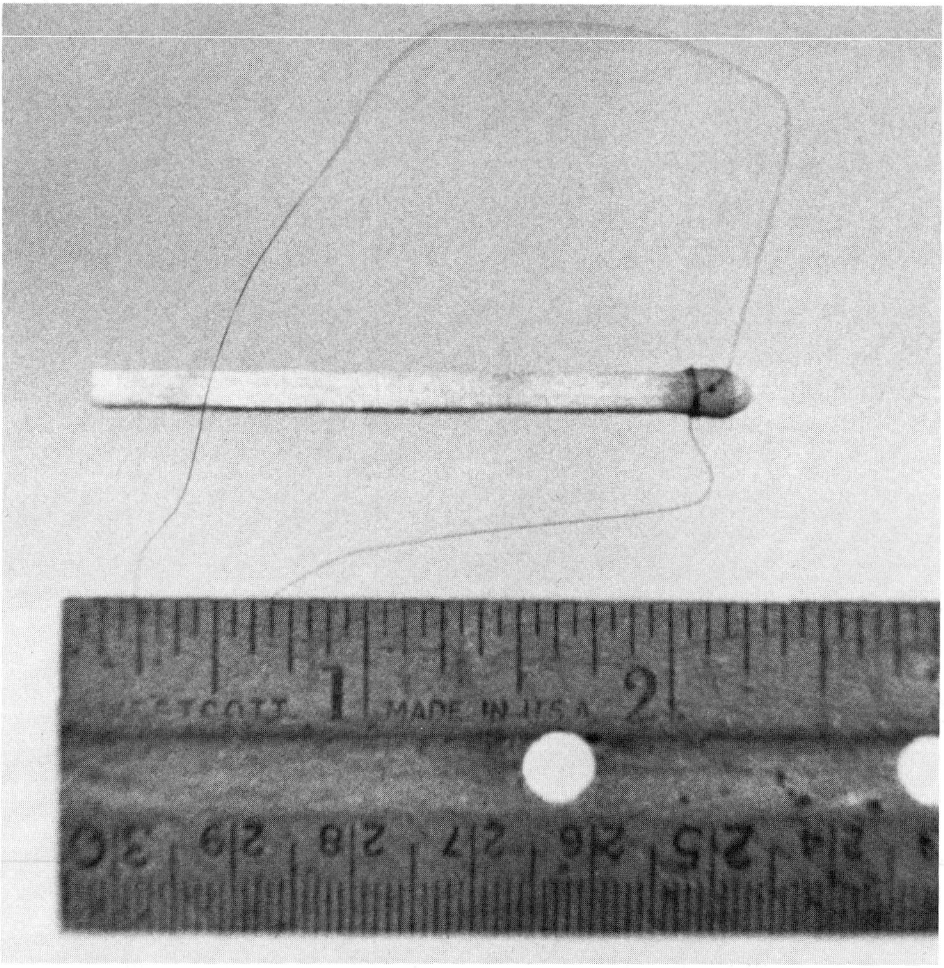

Figure 4-42.

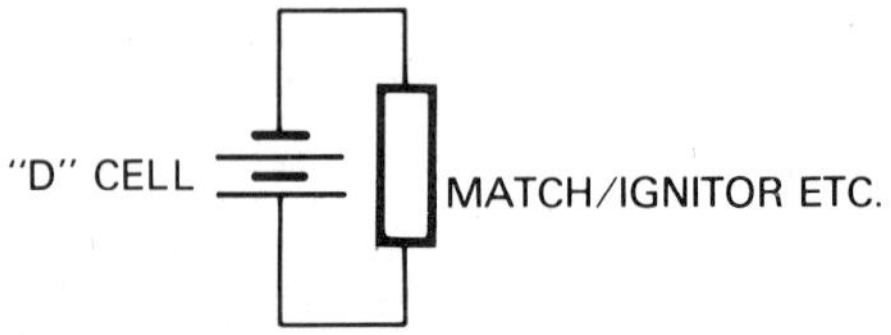

Figure 4-43.

Light Bulb Incendiary

This device creates an explosion, with subsequent ignition of nearby flammables, when the circuit to the light bulb is turned on.

The device is prepared by drilling a hole, about 1/8 inch in diameter, in the base of the bulb (Figure 4-46).

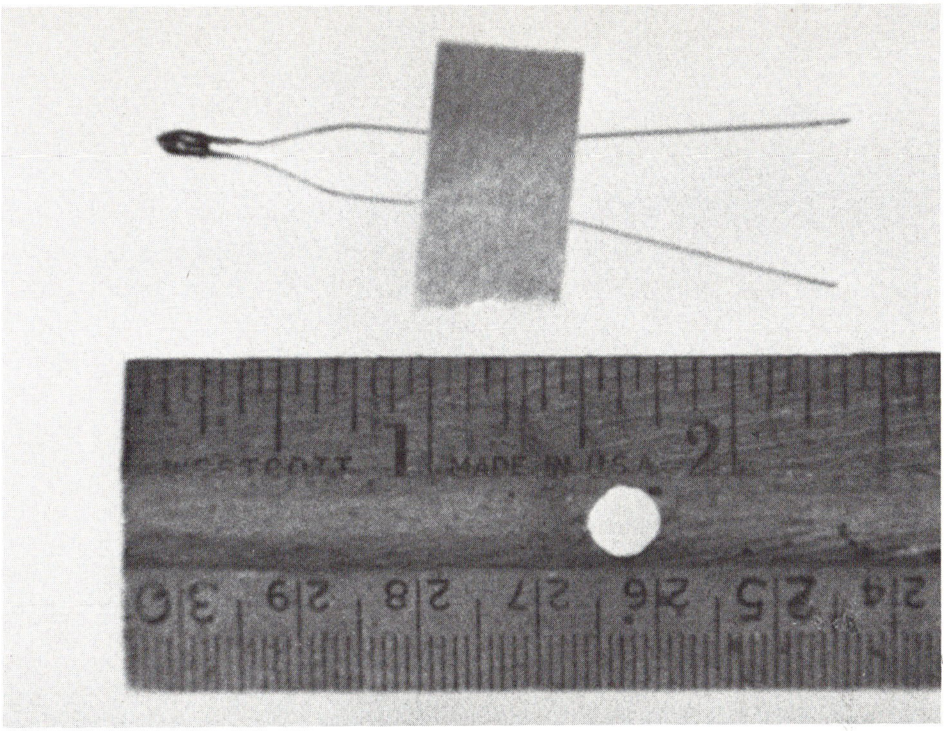

Figure 4-44.

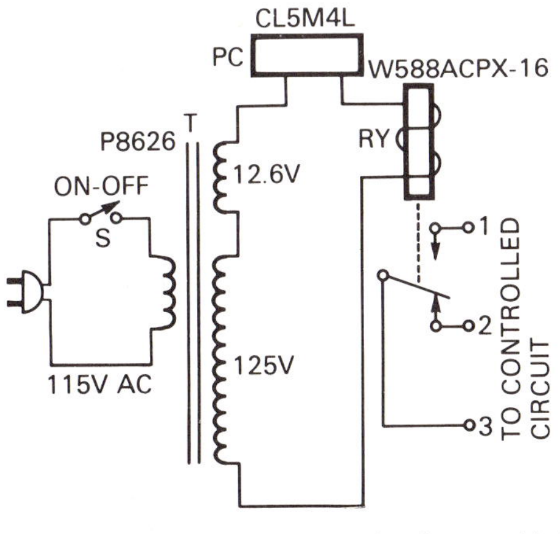

Figure 4-45. Turner: Solar Cells and Photocells. Howard W. Sams, p 44, 1975.

About three to six drops of gasoline are placed into the hole with an eye-dropper (Figure 4-47).

The hole is then sealed with a piece of electrician's or masking tape (Figure 4-48).

Before inserting the "device bulb" into the socket, a regular bulb is inserted and the circuit tested to be sure it is "dead." Turning on the circuit causes the explosion.

Figure 4-46.

Figure 4-47.

Nichrome Wire and Milk Cartons

Nichrome wire from a toaster, hot plate, electric heater, or other similar device, is the primary ingredient. The second ingredient is a plastic milk carton

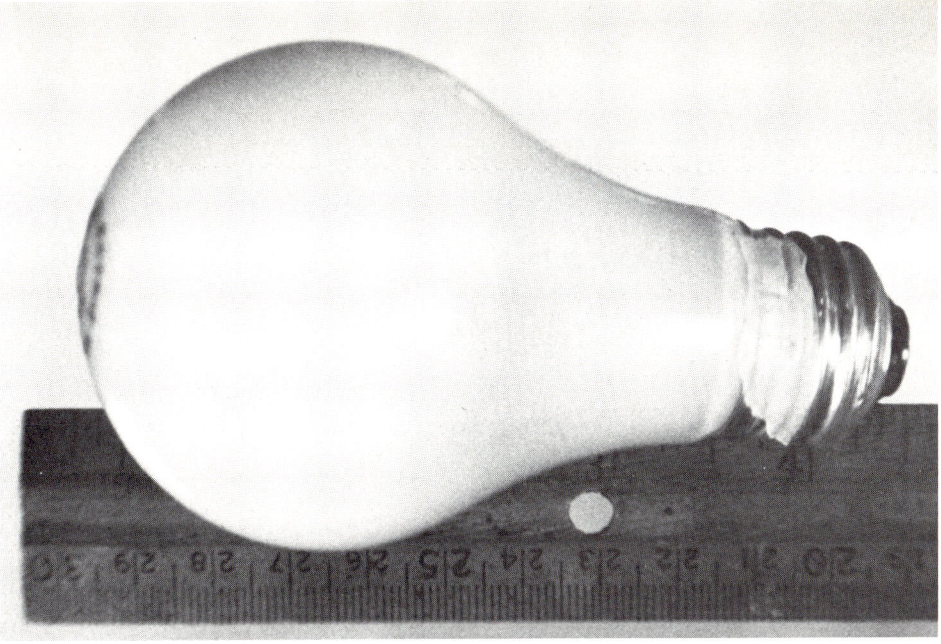

Figure 4-48.

filled with a flammable liquid. For this text the nichrome wire was removed from a hot plate (Figure 4-49).

The plastic container is tested to confirm that it will hold the flammable liquid without being destroyed by the flammable liquid.

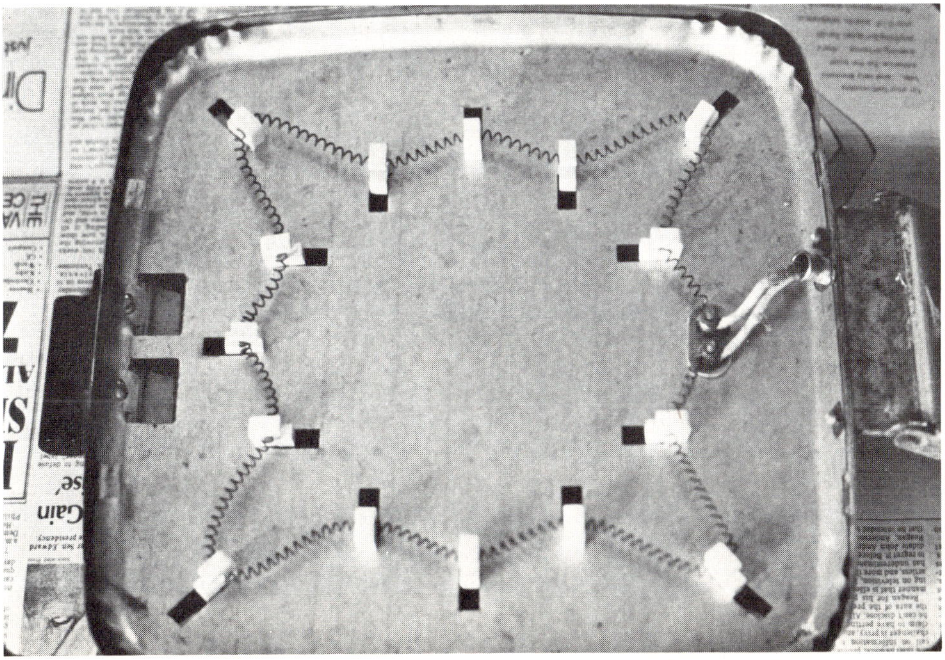

Figure 4-49.

Too long a length of wire may add too much resistance, which will result in insufficient power. The nichrome wire is then randomly taped around the exterior of the container, as shown in Figure 4-50.

The flammable liquid could be "straight" or a mixture of various flammables such as gasoline/turpentine, kerosene, or varsol.

The ends of the nichrome wire are affixed to a plug, or to their original bolts and nuts in the heater or hot plate. A timer could be used between the device and the a.c. receptacle. Heating the wire causes the plastic to melt, and the hot wire then ignites the flammable liquid.

Figure 4-50.

Televisions

The objective of all incendiary devices is to make the fire's origin appear accidental. Since televisions have various high and low voltages (some in excess of 20,000 volts), the opportunity exists to start a fire that could be attributed to a short circuit, electrical malfunction, or other 'obvious' cause.

To accomplish this end, the second anode is disconnected from the CRT (picture tube). This is done by first "grounding" the second anode to the chassis and then "unclipping" the lead from the hole in the CRT (see Figure 4-51).

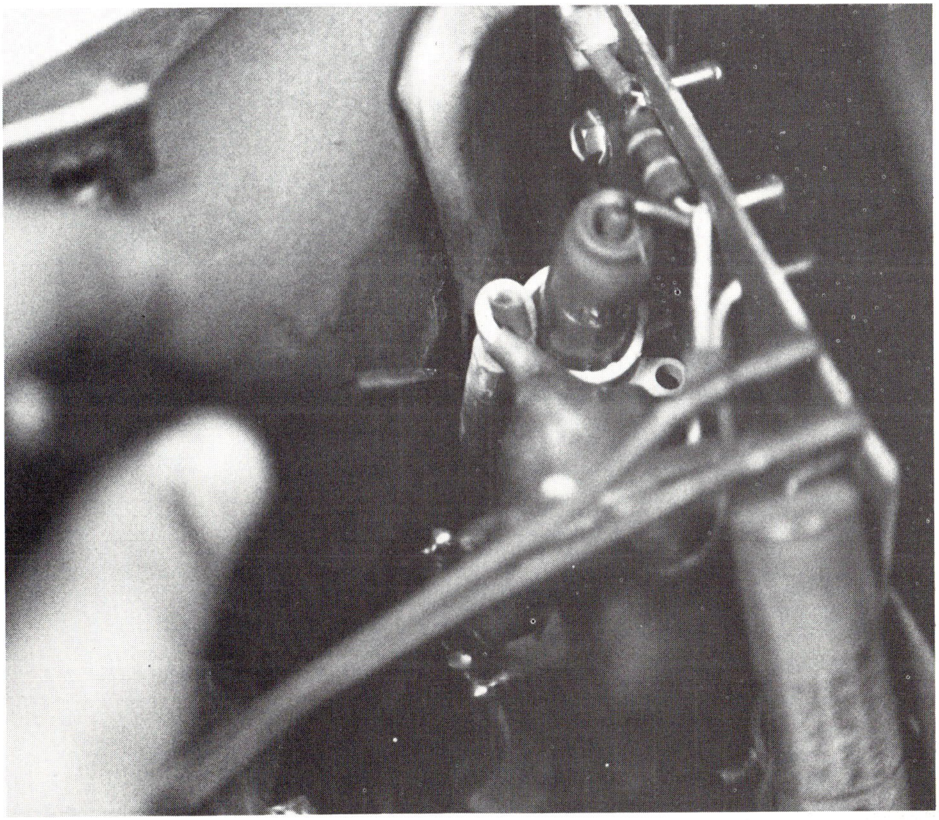

Figure 4-51.

Using a stiff wire or other substance, the second anode is then supported about ¼ to ½ inches away from the metal chassis (see Figure 4-52).

Flammable liquid is sprayed in and around the area of the "set up." Most of the tests that were successful for the author involved a volatile flammable liquid (gasoline) and a "heavier" flammable liquid (turpentine and/or fuel oil).

If the device is placed on a bookshelf, there would most probably be sufficient additional fuel present to guarantee success. If the fire is totally successful, the "evidence" (CRTs and other components) normally is destroyed.

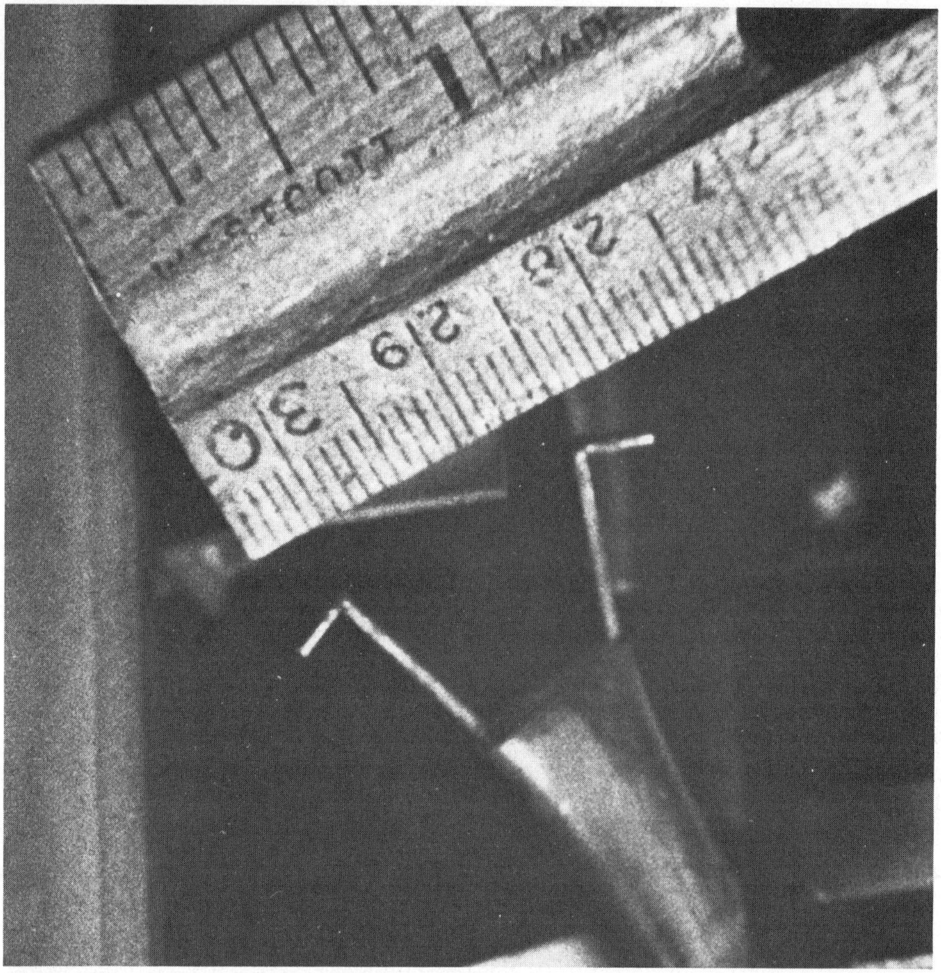

Figure 4-52.

Cartridge Fuses

Although it is difficult, if not impossible, to start a fire with any type of fuse or circuit breaker (overcurrent protection device), certain things can be altered in this safety device so that the protection device never "sees" the fault it is supposed to protect against.

Cartridge fuses can be by-passed effectively in a very simple fashion. The fuse is prepared by removing one of the end caps and cutting the fusible link (strip) inside the cap. (It may break while the disassembly process is in operation.) (Figure 4-53.)

At this point a nail is cut to a length just slightly longer than the length between the end caps. The nail is then inserted inside the cartridge and the end cap is replaced. The entire fuse is placed back in the panelboard.

An overload is introduced; however, the "fuse" will not see the fault and will allow the wire to overheat. Recent test results indicate that the potential for fire in walls as a result of overheating is most likely where cellulose insulation is involved and where about 30 amps can be drawn through a #14 solid copper wire for one hour or more.

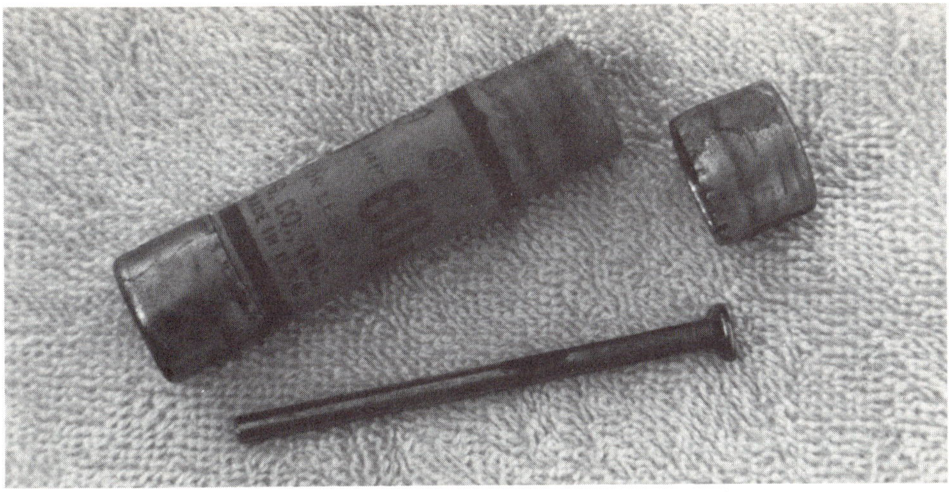

Figure 4-53.

Obviously, visual examination is not sufficient. All fuses should be checked by visual and electronic test equipment.

Circuit Breakers

As with cartridge fuses, circuit breakers can also be altered to change their rating. A typical circuit breaker is shown in Figure 4-54.

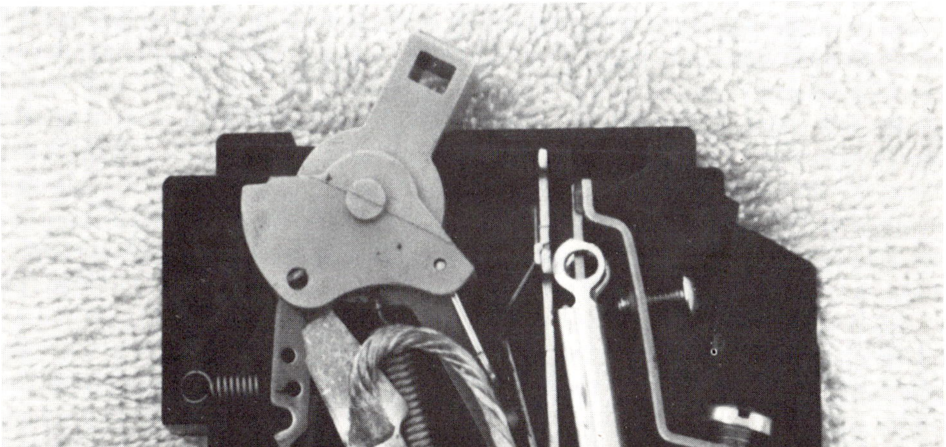

Figure 4-54.

Removing the plug along one edge of the circuit breaker exposes a screw head. This screw head sets the "trip" mechanism, which means it will alter the current required to trip the circuit breaker (see Figure 4-55).

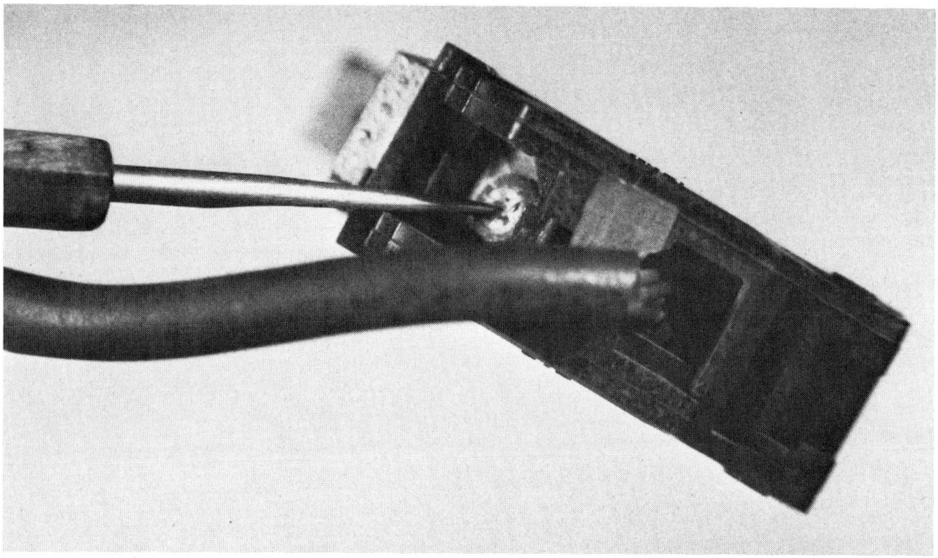

Figure 4-55.

Although no data is available to directly support this, turning the screw ¼-turn clockwise may increase the amperage rating by five amps. This would mean that one full turn would change the rating by approximately 20 amps. Two full turns would create a 60-amp breaker out of a 20-amp breaker.

The plug is replaced, if so desired, and the circuit breaker re-inserted in the panelboard. An excessive current draw (as described in Cartridge Fuses) is then developed, but the circuit breaker would never "see" the problem.

From an investigator's point of view, the only way to detect what was done is to thoroughly examine each and every circuit breaker. If the plug was replaced, visual examination will not be sufficient. It will also be necessary to have a laboratory test the circuit breaker to determine if it trips at the rated current draw, time, and temperature it was originally designed for.

Edison Base Fuses

Edison base fuses, like circuit breakers and cartridge fuses, can be tampered with in such a manner so that they do not electrically protect the circuit. An Edison base fuse is shown in Figure 4-56.

The fuse is altered by prying loose a small section of the copper exterior threaded section. The loosened piece is then bent over against the center conductor pin of the fuse. It may or may not be soldered. Figure 4-57 shows the completed device.

The by-passed fuse is then screwed back into its slot in the panelboard and the overload is introduced, as in the past two devices.

Figure 4-56.

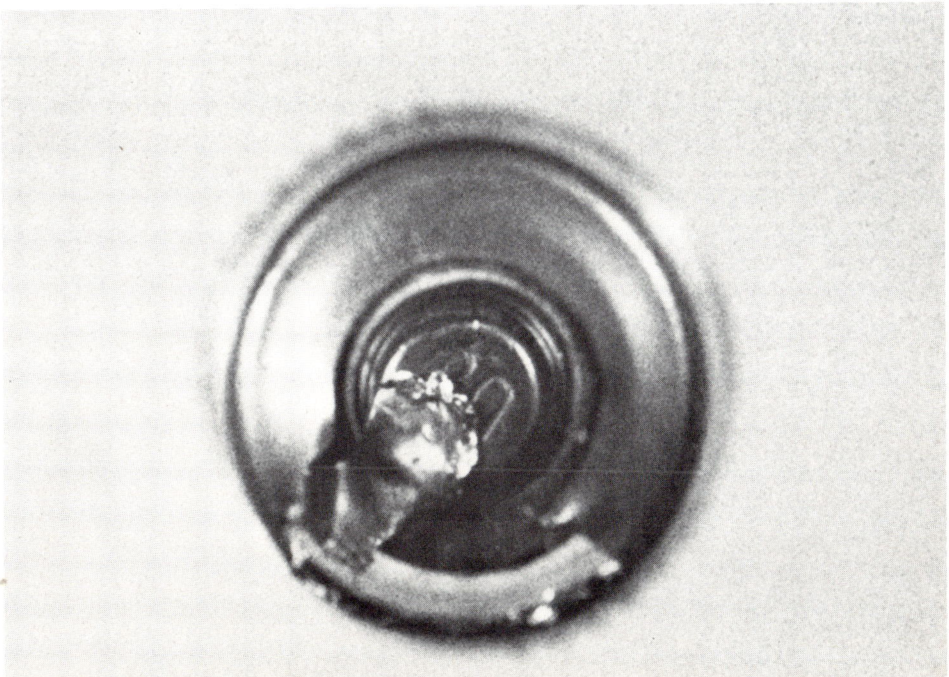

Figure 4-57.

Any investigator who "casually" looks at the fuses would see only the fuse "not blown." An adjacent fuse may be blown, depending on fire impingement on the wires of the circuit it controls. This device is usually successful because detection means checking each and every fuse, both visually and with an ohmmeter.

Hygrometric Triggering

Hygrometric triggering is an unusual mechanism. The sensing of moisture completes an electronic circuit, which, in turn, can activate a relay or other initiating device. Figure 4-58 is a schematic of such a hygrometric device.

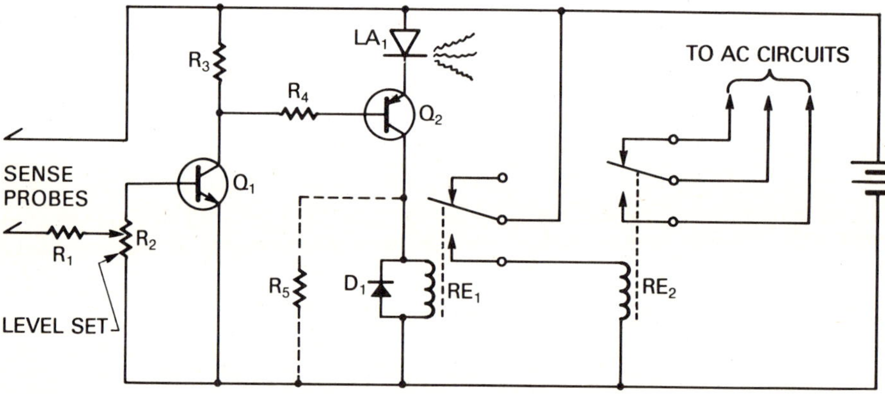

Figure 4-58. Moisture Sensing Switches, "Information Unlimited", Box 716, Amherst, NH, Reprinted by permission.

Theoretically the circuit operates as follows. The sensing probes provide a current between themselves. Resistor R3 determines the current at the base of transistor Q2 for any given current between the sensing probes. When Q2 conducts, a negative voltage will bias Q1 into conduction, which then causes the relay to operate.

R3 is adjusted "on-scene" to trigger a test circuit with whatever percentage of moisture is available or desired. The test circuit is then removed and the live initiating device is attached to the normally open relay contacts. A nine-volt battery will provide the necessary voltage to operate the circuit.

Proximity Triggering

This type of trigger would most likely be set up in a house or building equipped with a perimeter proximity alarm system. Instead of setting of the alarm system, however, it triggers the ignition device.

The absence of a perimeter alarm system does not preclude the use of this type of triggering, providing there is time to set up the system and test it. The technical considerations normally preclude the use of this system in a "one-shot" affair.

The principle of this system involves two wires acting as a capacitor. One wire (the ground) acts as one plate of the capacitor while the other (the sensing) acts as the chassis plate. As long as the charge is "static" (non-changing), the capacitance is constant. As soon as any object approaches the wires, the capacitance between the two wires changes, thus changing the voltage factor, which ultimately varies the current. A schematic for a proximity trigger is as follows:

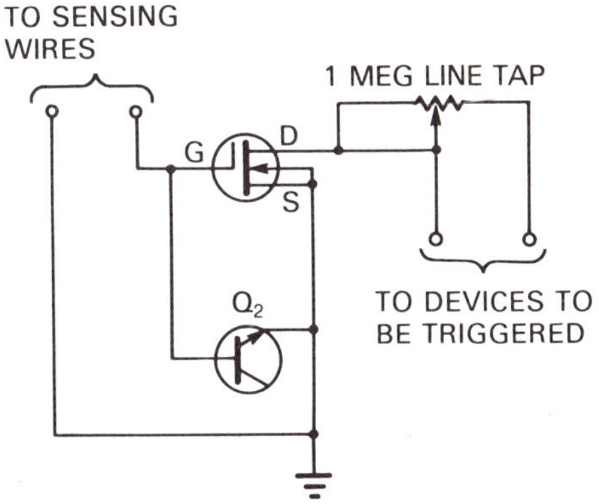

Figure 4-59.

Transistor Q1 is an n-channel MOSFET transistor similar to a Motorola® HEP-F2005 or an ECG 220. Q2 is an NPN Silicon transistor such as a WEP 735A, ECG 123A, Archer® 276-2016, HEP 54 or Archer® 276-2009.

Usually this will be set up where a fence or deck rail can carry the wires. A problem with this arrangement is that sensitivity decreases in outdoor applications. There is also the possibility of interference from nearby power lines.

A secondary method is to string the lines around the main door frame. A person passing through the door causes the necessary capacitance changes between the wires. From an analyst's viewpoint, the presence of wires around the interior of the door frame or on fence or deck railings should be a signal to search further and inquire as to the existence of a perimeter alarm system.

For analytical purposes, two additional circuits, considered to be proximity switching circuits, are also shown.

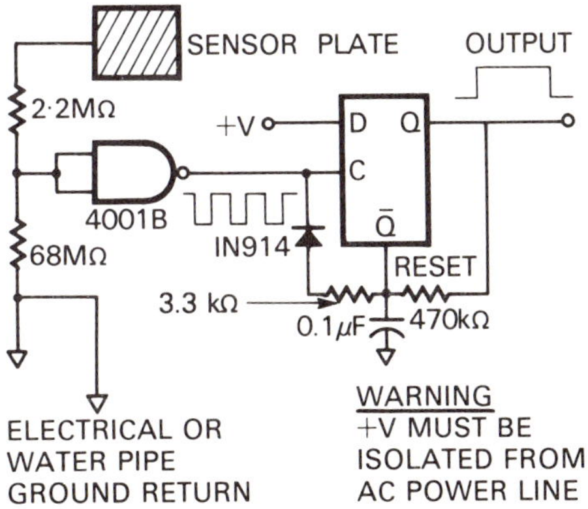

Figure 4-60. PROXIMITY SWITCH—Based on coupling of human body to 60-Hz power line. Hand held close to sensor plate induces hum into 4001B gate. This is squared and used to trip 4013 retriggerable mono MVBR. Output is clean from instant of first proximity until several milliseconds after release. Sensitivity depends on size of plate.—D. Lancaster, Clocked Logic, *Kilobaud*, May 1977, p 24-30.

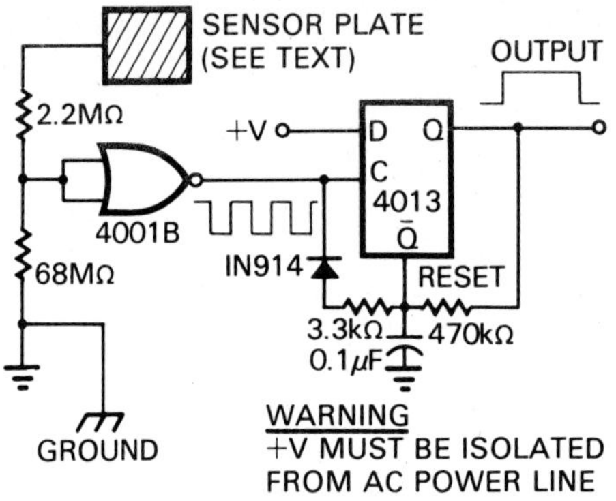

Figure 4-61. PROXIMITY SWITCH—Hand brought near sensor plate induces 60-Hz power-line hum in section of quad two-input NOR gate. Hum is squared by gate and used to trip section of 4013 connected as retriggerable mono MVBR. Output of mono is clean from instant of first proximity until several milliseconds after moving hand away. Sensitivity depends on size of metal plate and number of permissible false alarms from other noise sources nearby.—D. Lancaster, "CMOS Cookbook," Howard W. Sams, Indianapolis, IN, 1977, p 278-282.

Proximity triggering can also be used with burglar alarms. The basis for use with a burglar alarm system is that breaking a circuit (stone through window, motion detector sensor, noise-activated trigger from alarm bell ringing, etc.) activates the ignition device, or that a normally open circuit activates upon receipt of a signal (outgoing call being placed by automatic dialer tape mechanism, etc.).

Usually there are three parts to any alarm system: (a) sensor; (b) control unit; and (c) annunciator or reporting device. The sensors (or detectors) operate on the principle of "touch" through the use of magnetic, spring, micro, or mercury switches, pressure sensitive pads or mats, metal contact (foil tapes), lacing wires through screens or grids, trip wires, or vibration detectors.

A second type of sensor operates on audio principles, most particularly ultrasonic (doppler), or through microphone pick-ups such as "hot-wired mikes" on telephones.

The third principle of sensors uses light or sight sensing like photoelectric beams, light intensity changes in circuits, and closed circuit television systems.

Control units represent the "brain" of the system and usually consist of a power supply, time delay circuit, anti-tamper devices, and intrusion signal transmission system. The latter system could be directly wired to the telephone lines, microwave systems, or radio links.

Many control systems have internal (user) check systems so the owner can check the system. Some control units are designed to monitor several circuits simultaneously. Some units monitor, determine whether the signal being received is for police or fire, and then make the appropriate outgoing call.

The annunciator or reporting device is simply the end result of the system and includes bells, buzzers, horns, lights, and other similar devices for notification of the owner (local alarm) and of a central station (remote alarm).

Alarm circuitry, in its simplest state, consists of a normally open or normally closed circuit. As long as the circuit remains open, no alarm is given in the case of the normally open circuit. For the normally closed circuit, an alarm is sounded if the circuit is opened.

More complex circuitry becomes involved when moving to tilt and vibration switches, plunger switches, taut-wire and spring systems, modulated light beams, light beams deflected by mirrors, infrared bodyheat detectors, and ultrasonic wave detectors.

Since a working knowledge of these systems is required, the property owner's knowledge of electronics, and the most recent service call by the alarm service company should be determined. To determine if the system is "normal," see if the owner control switch is set in the proper position. This is usually a key control; nomenclature may include terms like "Power Off," "Test/Reset" (no alarm will be given), "Sentinel" (standby position), "Delayed Alarm" (user has x number of seconds/minutes to turn off system before alarm is given), and "Instant Alarm" (no delay from time circuit is broken or completed until time alarm is given).

Beyond this, however, there could be additional circuits that should be checked inside the "box" (for which the customer may have a key).

Some units have a service personnel switch, such that, even if the customer switches are in the correct "Alarm" position, no alarm will be given. In one system, this "Service/Maintenance Switch" (normally not accessible to the owner) can be positioned to turn off the dialer system. Another position allows the dialer to run through its sequence (appearing to be normal); however, monitoring will reveal nothing going out over the phone lines!

There are also alarm systems that can be "rigged" simply by locating the phone line pair from the building, and placing a clip-lead across the pair. The system 'sees' a complete circuit. Phone company installers may even write the coding for the pair on the wall cover.

The telephone company may also require installation of a line coupler or line seizure device. Some alarm systems have a built-in or "additional charge purchase" seizure, which will seize the line to place an outgoing call even though the line is in use. The seizure device could also work in the event of the telephone lines being shorted out by flame, depending on the location of the fire versus the line seizure unit. By-passing this device, or removing it from the circuit, is always a possibility so check whether or not this device existed.

For reference purposes, a number of burglar alarm circuits are presented (See Figures 4-62 through 4-85).

Bimetallic Triggering

Dissimilar metals, with unequal coefficients of expansion rates, are used in this simple but effective device. As one metal heats (or cools), it will move toward (or away from) the opposite piece of metal, which is of a different metallic structure. This principle causes a circuit to be made or broken. The most common examples of this principle are found in temperature controls for thermostats, coffee pots, electric fry pans, water heaters, clothes dryers, and other similar appliances. Figure 4-86 shows the bimetallic strip and circuit in the open position. Figure 4-87 shows the circuit complete after the bimetallic strip has warped due to the presence of heat or cold.

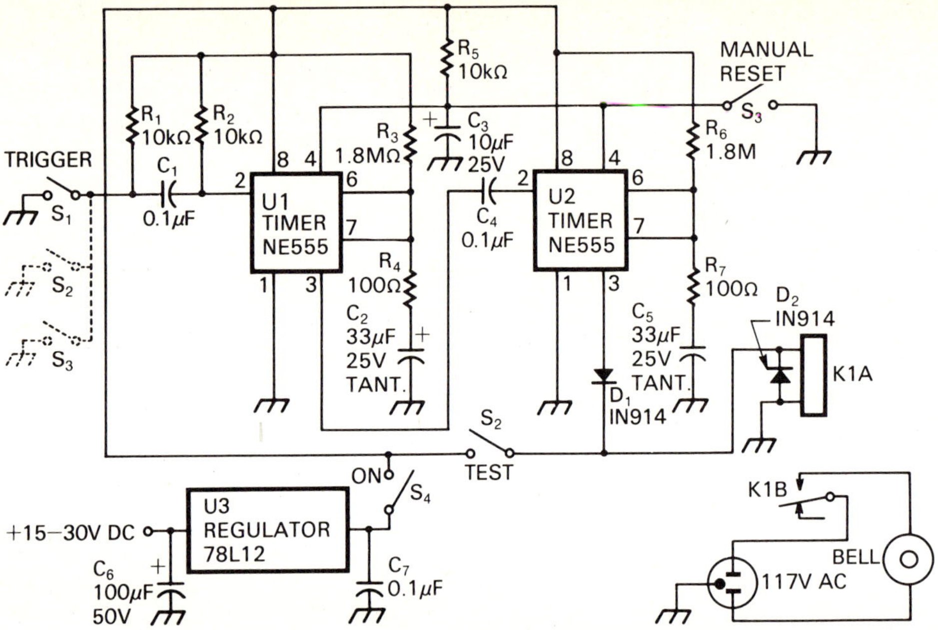

Figure 4-62. ENTRY-DELAY ALARM—First 555 timer provides delay of about 20 s after triggering by sensor before alarm bell is energized, to allow thief to be caught inside house or give owner time to enter and shut off alarm. Alarm then rings for about 60 s under control of timer U2. Alarm period was set short to attract attention without unduly annoying neighbors.—J. D. Arnold, A Low-Cost Burglar Alarm for Home or Car. *QST*, June 1978, p 35-36.

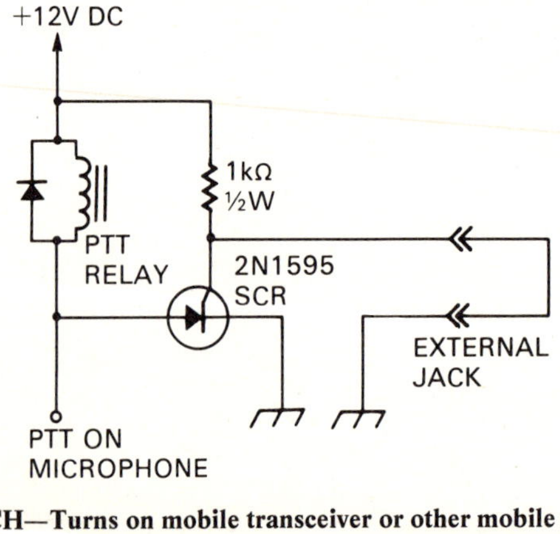

Figure 4-63. SCR LATCH—Turns on mobile transceiver or other mobile equipment when power is applied, if external circuit is broken when equipment is stolen. Transmitter will then put unmodulated carrier on air even with PTT switch disconnected or off, for tracing with radio direction finder. If added components are carefully concealed in equipment and new external wiring is worked into existing wiring harness, few thieves will be able to locate trouble. External wires are run under dash so thief must cut them to get out equipment. PTT relay should have protective diode. SCR is 100 PIV, 1 A, but HEP R1003 or R1217 can also be used.—E. Noll, Circuits and Techniques, *Ham Radio*, April 1976, p40-43.

Polarity of the bimetallic strip is unimportant, so whatever polarity is needed for the ignition device will be the governing factor. As used from Figures 4-86 and 4-87, the "output" could be directly connected to a relay providing

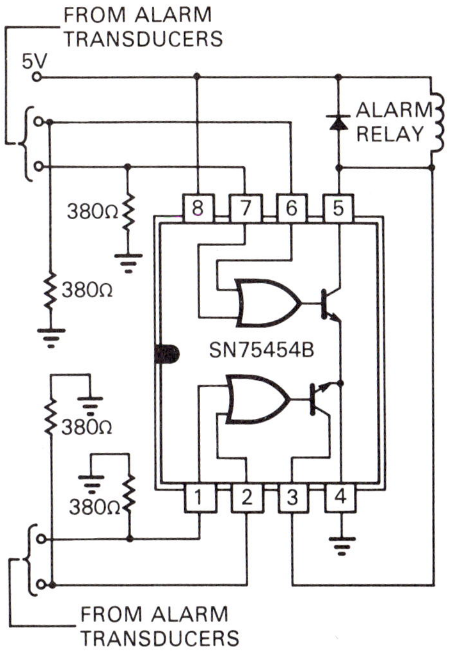

Figure 4-64. ALARM-SIGNAL DETECTOR—Texas Instruments SN754548 dual peripheral positive-NOR driver energizes alarm relay when alarm signal is received from any one of four different alarm transducers.—"The Linear and Interface Circuits Data Book for Design Engineers," Texas Instruments, Dallas, TX, 1973, p 10-66.

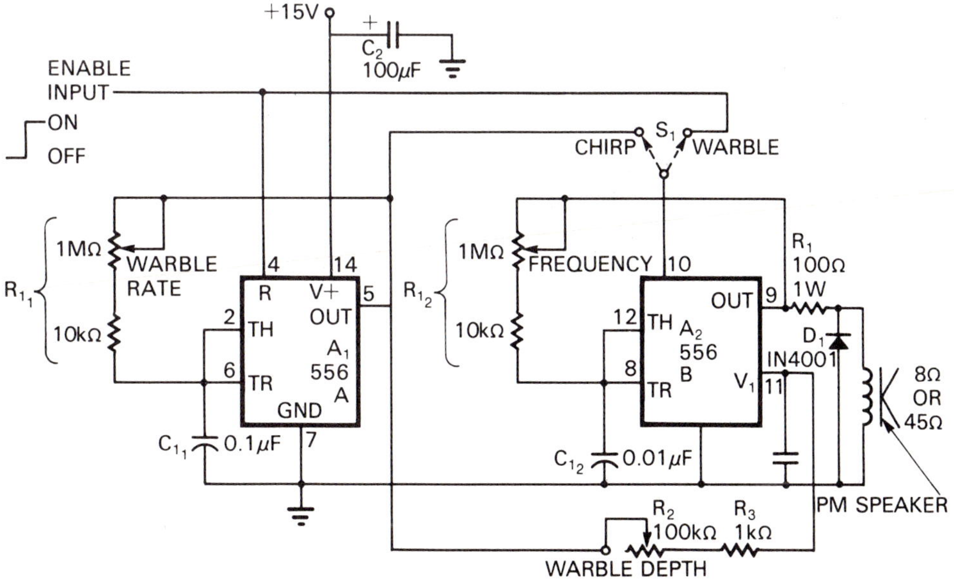

Figure 4-65. AURAL INDICATOR—Provides attention-getting chirp sound, warble, or continuous tone when turned on by high input from burglar-alarm sensor circuit. Second section of 556 timer provides optional frequency modulation of basic tone to give warbling effect. Chirp is achieved by gating tone oscillator on only during high states of warble oscillator. Aural sensitivity is maximum in range of 1-2 kHz, set by value of R_2.—W. G. Jung, "IC Timer Cookbook," Howard W. Sams, Indianapolis, IN, 1977, p 232-235.

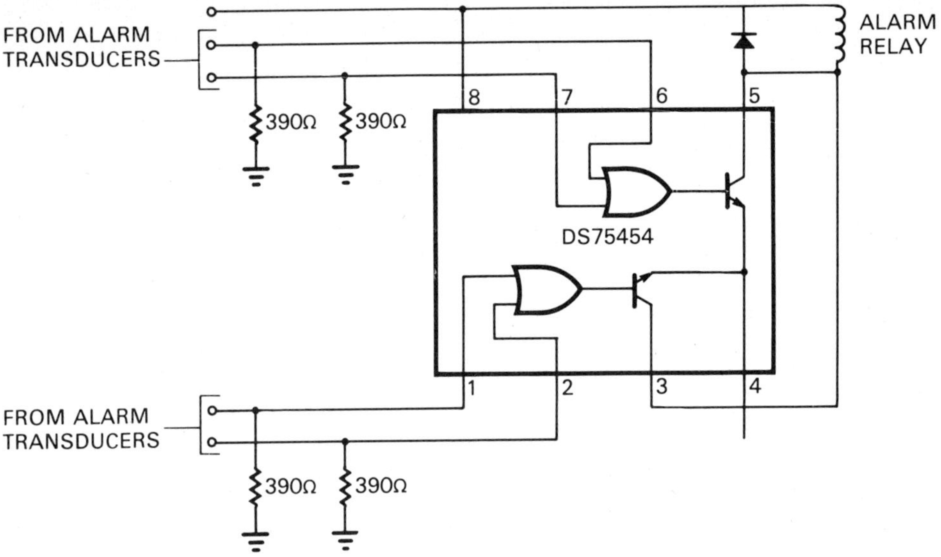

Figure 4-66. ALARM DETECTOR—National DS75454 dual peripheral NOR driver operating from single 5-V supply energizes alarm relay when one of alarm transducers for either section delivers logic signal as result of intruder action.—"Interface Databook," National Semiconductor, Santa Clara, CA, 1978, p 3-20-3-30.

there is sufficient voltage and current to drive the relay from the power source. The odds on finding this small device in any fire debris are rather remote!!

Remote RF Triggering

Remote RF triggering is accomplished through the use of two-way radios. The most available, and perhaps most inexpensive, would be CB transceivers. Unfortunately, they are limited in power and range, so the next most logical choice would be something in the mobile telephone or public service bands (150-170 MHz).

Use of this type of triggering requires a sending and receiving unit. The receiving unit is placed at the scene while the sending (transmitting) unit stays with the person who will trigger the device.

Receipt of a signal by the "on-scene" receiver activates the triggering device. Unfortunately, especially with CBs, with so many units in use at any given time, the device could be triggered prematurely (even while it is being set up!).

To eliminate this potential, "code-tones" are inserted in the audio processing chain of the transmitter. This means that, in addition to the RF carrier frequency, there must also be a special tone (or group of tones) present or the receiver won't "see" the incoming RF signal. This is accomplished by inserting tones from an encoder at the transmitter, which are decoded at the receiver.

The simplest way to obtain an encoder for the transmitting portion of the operation is from the tone pad of a telephone. Figure 4-88 shows a schematic of the outboard components needed with such a pad and Figure 4-89 shows a unit ready for operation.

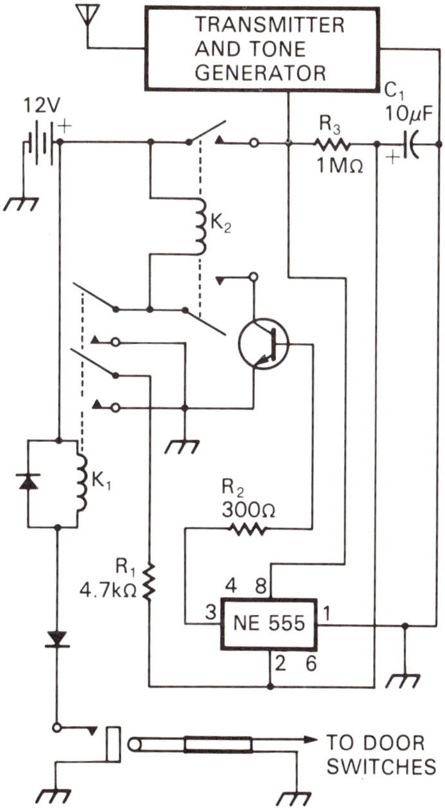

Figure 4-67. SILENT ALARM—When thief opens car door, relays K_1 and K_2 activate tone-modulated transmitter, which can be any legal combination of power, frequency, and antenna. A few milliwatts of power should be adequate. Thief hears nothing, but owner is alerted via portable receiver tuned to transmitter frequency. Transmitter remains on about 15 s (determined by R_3 and C_1) after door is closed until NE555 times out and removes power from transistor. Use any NPN transistor having adequate current rating for relay. If alarm is provided with its own battery and whip antenna, it can not be disabled from outside of car.—A. Day, Soundless Mobile Alarm, *CQ*, April 1977, p 11.

The decoder at the receiver must be capable of detecting the tones being sent. This is accomplished through the use of an NE567V Phase Lock Loop integrated circuit chip. If a telephone tone pad was used as the encoder source, then two chips would be needed in the decoder circuit. One chip detects the "high" tones while the other chip detects the "low" tones.

When both of the PLL's receive a signal, their outputs will go "low," which will change the "state" of IC3 to a logic "high." This, in turn, will cause the transistor to conduct, which activates the relay. As in previous circuits, the "load" or "ignition device" is attached to the appropriate relay contacts. Figure 4-90 is a schematic of the decoder section of the receiver portion of the device.

The encoder signal is connected in parallel with the microphone input to the transmitter and the appropriate button is pressed while the push-to-talk (microphone keying button) is depressed.

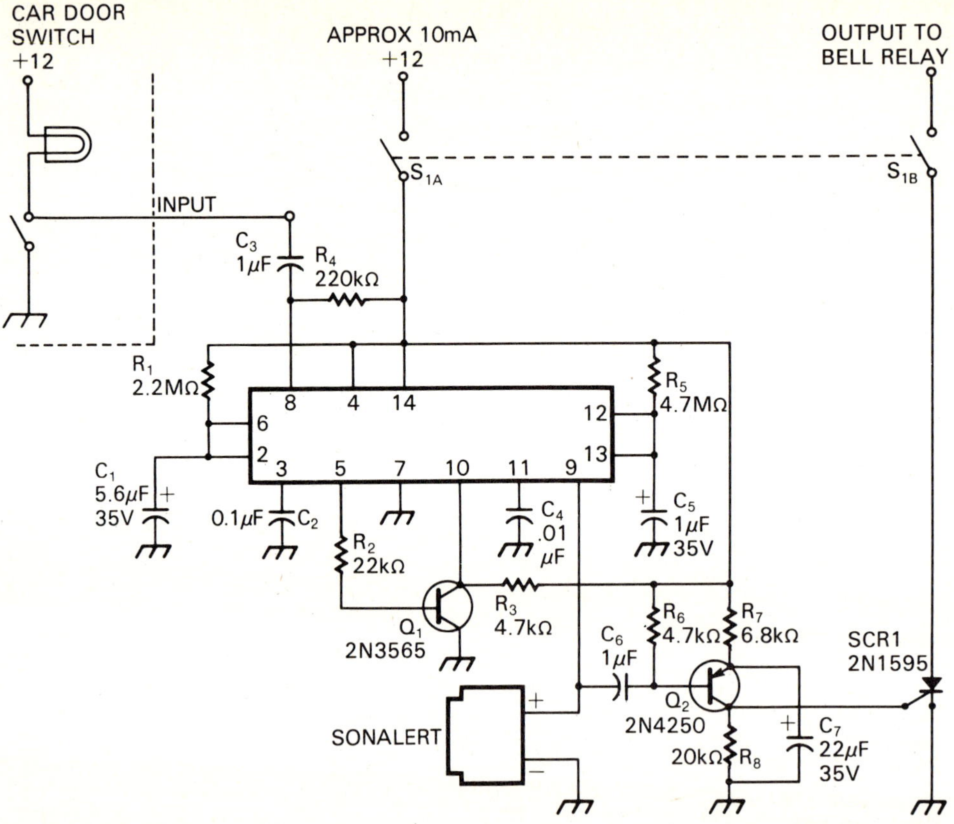

Figure 4-68. CAR-THEFT ALARM—Alarm remains on even if signal from car door switch or other sensor is only momentary, so relay is wired to be self-latching until keyswitch S1 is turned off. Use hood locks or hood-opening sensors to prevent thief from disabling alarm by cutting battery cable. Circuit includes time delay of 6 s for entering car and shutting off alarm, to avoid need for external keyswitch. Sonalert makes loud tone during 6-s delay period to remind driver that alarm needs to be turned off. At end of 6 s, Sonalert stops and much louder bell is energized to further discourage intruder.—J. Pawlan, The Smart Alarm, *73 Magazine*, June 1975, p 37-41.

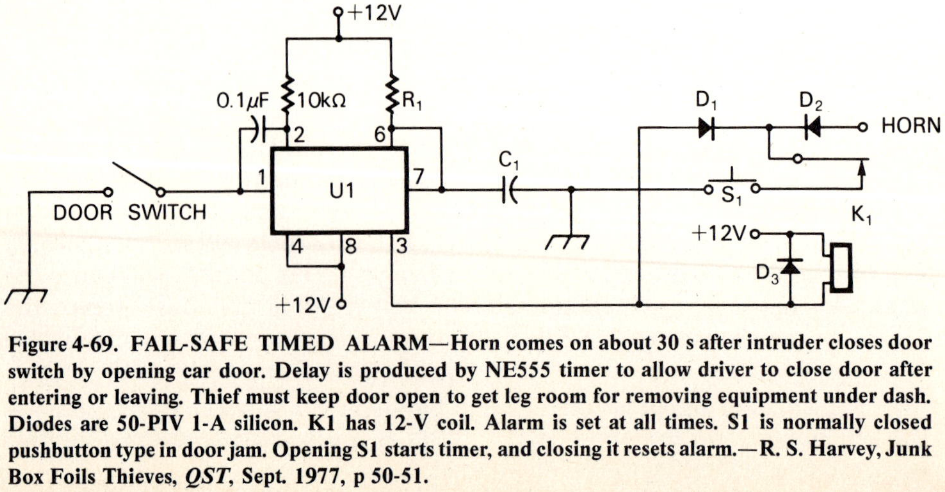

Figure 4-69. FAIL-SAFE TIMED ALARM—Horn comes on about 30 s after intruder closes door switch by opening car door. Delay is produced by NE555 timer to allow driver to close door after entering or leaving. Thief must keep door open to get leg room for removing equipment under dash. Diodes are 50-PIV 1-A silicon. K1 has 12-V coil. Alarm is set at all times. S1 is normally closed pushbutton type in door jam. Opening S1 starts timer, and closing it resets alarm.—R. S. Harvey, Junk Box Foils Thieves, *QST*, Sept. 1977, p 50-51.

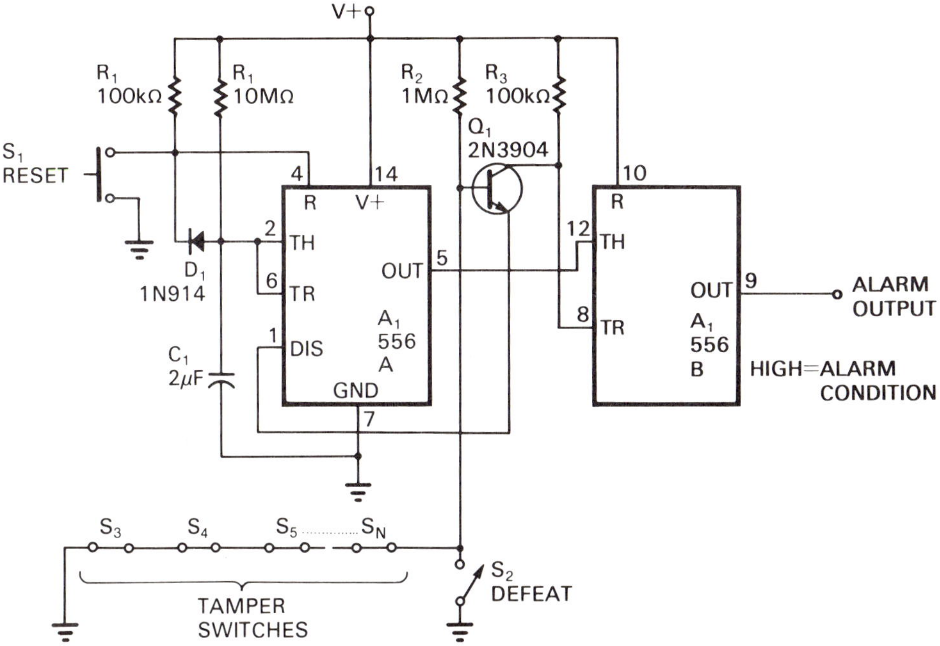

Figure 4-70. WINDOW-FOIL ALARM—Combination of power-up mono MVBR and latch, using both sections of 556 timer, drives output line high when sensor circuit is opened at door or window switch or by breaking foil on glass. Once alarm is triggered, reclosing of sensor has no effect: S_1 must be closed momentarily after restoring sensor circuit to turn alarm off. Circuit includes 22-s power-up delay that prevents triggering of alarm when it is first turned on.—W. G. Jung, "IC Timer Cookbook." Howard W. Sams, Indianapolis, IN, 1977, p 231-232.

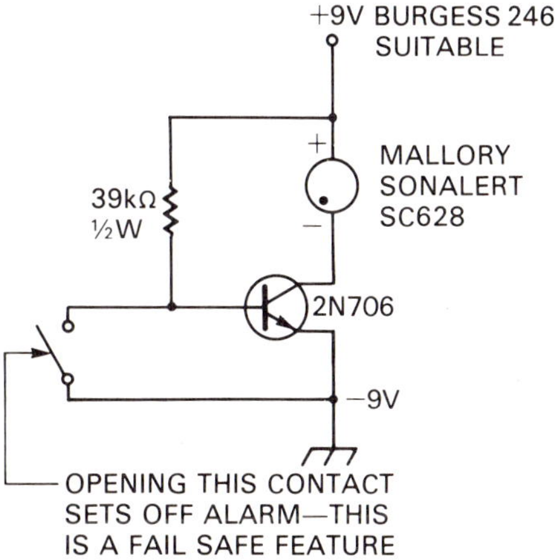

Figure 4-71. CIRCUIT-BREAKING ALARM—Operates from small 9-V battery, making it independent of AC power failure. Opening of switch or equivalent breaking of foil conductor removes ground from base of transistor, to energize alarm.—Circuits, *73 Magazine*, April 1973, p 132.

The decoder portion of the system is connected across the speaker leads or to the earphone jack of the receiving unit. The tone pad uses a combination of high and low tones for each number or letter. The chart in Figure 4-91 lists the frequencies.

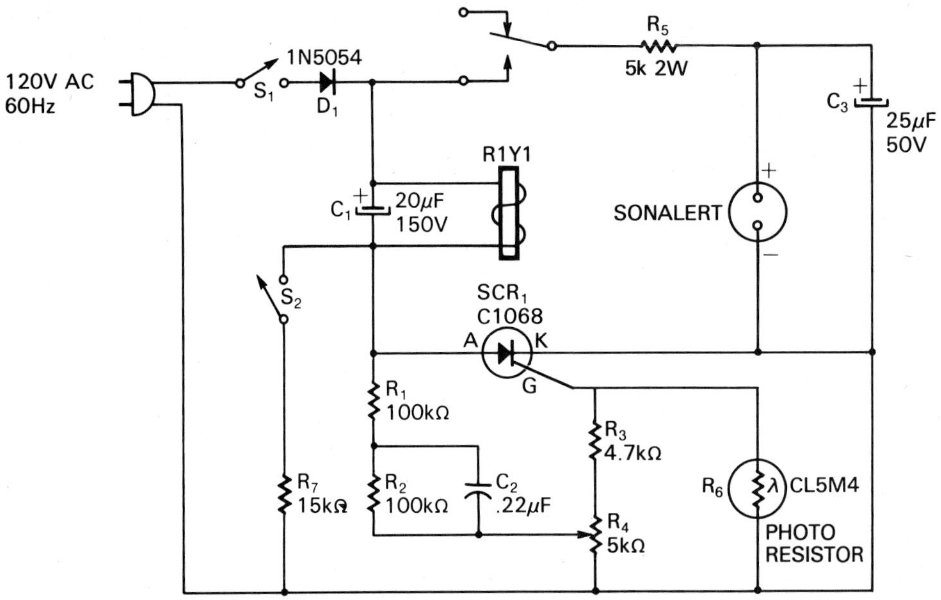

Figure 4-72. LIGHT-INTERRUPTION DETECTOR—Use of SCR as regenerative amplifier rather than as switch gives extremely high sensitivity to very slight reductions in light reaching photoresistor. Requires no light source or accurately aligned light-beam optics. In typical application as burglar alarm, light shining through window from streetlight provides sufficient ambient illumination so any movement of intruder within 10 feet of unit will energize Sonalert alarm. Sensitivity control R4 is adjusted so SCR receives positive pulses from AC line, but their amplitude is not quite enough to start regenerative action of SCR. Reduction in light then increases resistance of photoresistor enough to raise level of gate pulses for SCR, starting regenerative amplification that energizes relay. Use Mallory SC-628P Sonalert which produces pulsed 2500-Hz sound. With S2 open, alarm stops when changes in light cease. With S2 closed, alarm is latched on and S1 must be opened to stop sound.—R. F. Graf and G. J. Whalen, "The Build-it Book of Safety Electronics," Howard W. Sams, Indianapolis, IN, 1976, p 7-12.

Some additional remote control circuits are given as reference. Note that any remote control system could use wires, light beams, or be wireless. For additional references on possibilities, data transmission, instrumentation, optoelectronic, and radio repeater circuits should be studied (See Figures 4-92 through 4-96).

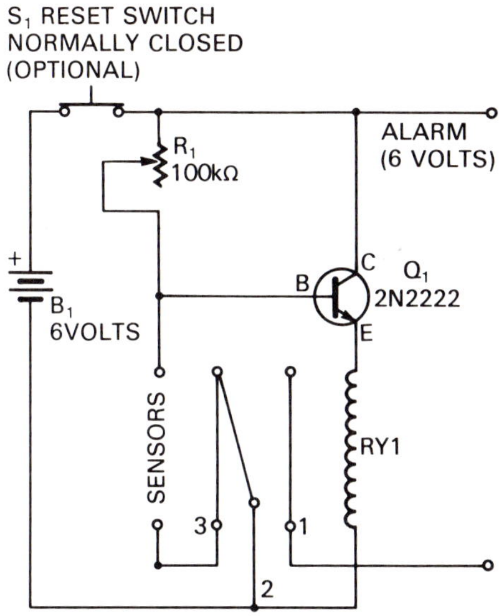

Figure 4-73. LATCHING ALARM—Closed-circuit alarm drawing only 130 μA of standby current from battery is turned on by opening sensor switch or cutting wire. Automatic latching contacts on relay prevent burglar or intruder from deactivating alarm by resetting sensor switch. Relay is Radio Shack 275-004. Sensor can be foil strip around window subject to breakage.—F. M. Mims, "Transistor Projects, Vol. 3," Radio Shack, Fort Worth, TX, 1975, p 75-86.

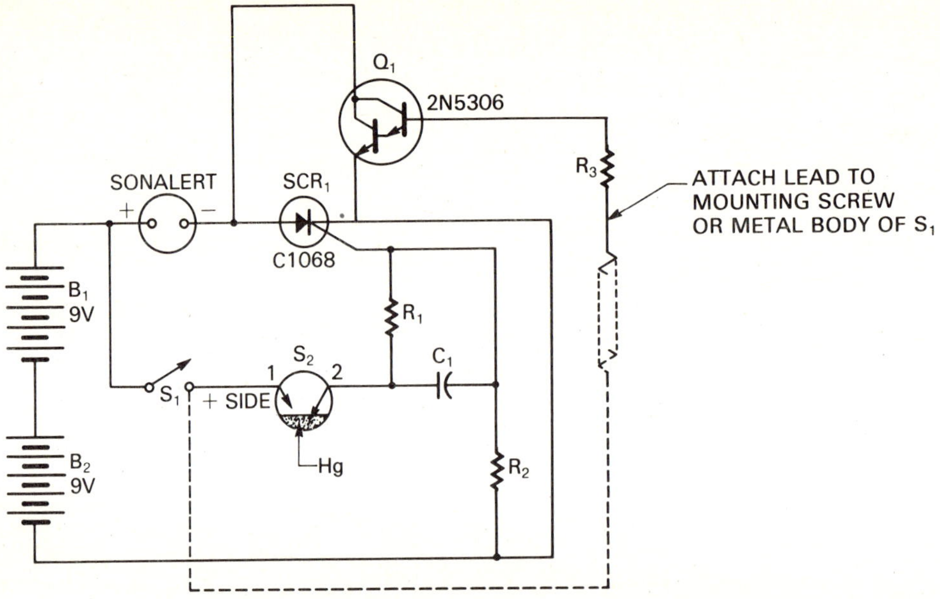

Figure 4-74. HOTEL-ROOM ALARM—Alarm mounted in flashlight-shaped cylinder is positioned on floor inside hotel room in such a way that it is knocked over by intruder opening door. Mercury switch S2 then triggers SCR and activates Mallory SC-628P pulsed Sonalert alarm. Circuit latches on and can be turned off only by use of Darlington-amplifier touch switch. Connection from base of Darlington to positive terminal of battery must be made through fingertips as shown by dashed line in order to silence alarm. Once silenced, S1 can be opened to disconnect latch so alarm can be moved. Other applications include protection of unattended luggage. C1 is 0.1 μF, R1 is 1 megohm, R2 is 1K, R3 is 39K, and S2 is mercury element removed from GE mercury toggle switch.—R. F. Gral and G. J. Whalen, "The Build-it Book of Safety Electronics," Howard W. Sams, Indianapolis, IN, 1976, p 19-24.

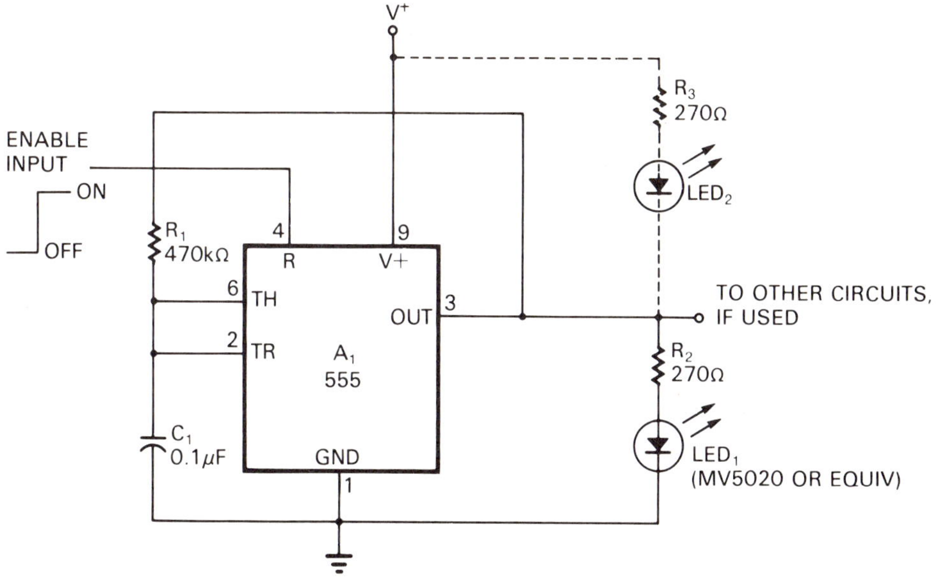

Figure 4-75. VISUAL INDICATOR—When circuit is activated by high output of burglar alarm circuit, 555 timer operating as very low frequency MVBR makes LED_1 flash on and off during alarm condition. Alternate connection of LED_1 to V+ holds LED_1 on for standby while flashing it during alarm. Oscillator output is also available for other uses if desired. Indicator can be located remotely from alarm.—W. G. Jung, "IC Timer Cookbook," Howard W. Sams, Indianapolis, IN, 1977, p 232-235.

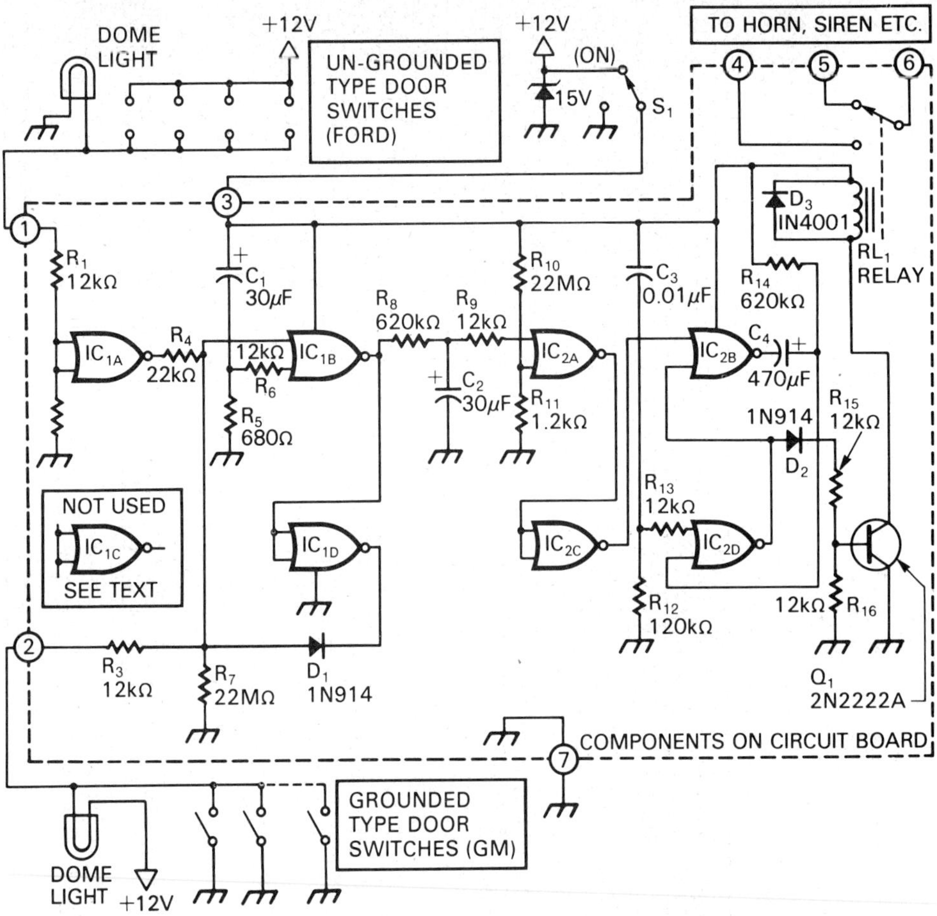

Figure 4-76. 5-min SHUTOFF—Vehicle intrusion alarm shuts off automatically in about 5 min after being triggered, as required by law in some states. Drain on battery is negligible until alarm is set off by intruder. Once triggered, operation sequence is not affected by subsequent opening or closing of doors. System uses two CMOS CD4001AE quad two-input NOR gates for switching logic. IC1 provides sensor interface, latch, and entry/exit time delays. IC2 provides output through Q1 and relay, as well as automatic shutoff delay. Article gives construction details and layout for printed-circuit board.—W. J. Prudhomme, Vehicle Security Systems, *73 Magazine*, Oct. 1977, p 122-125.

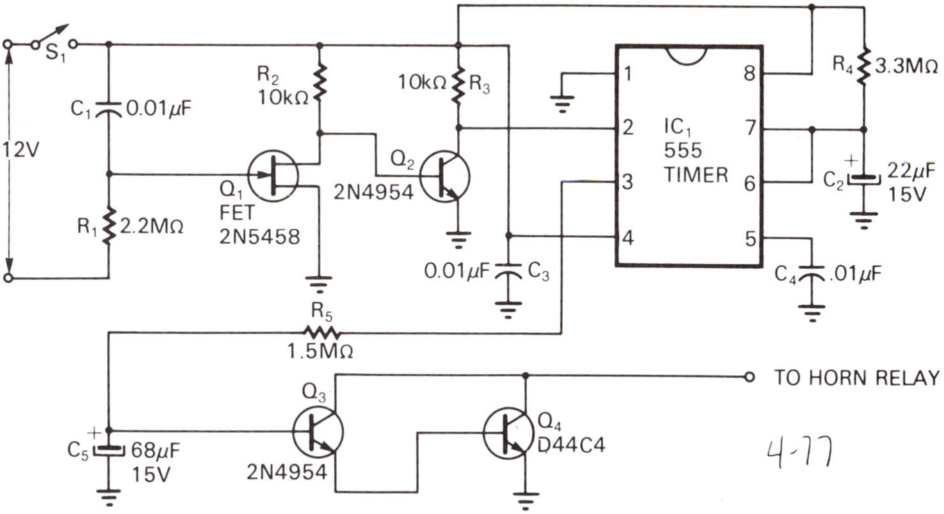

Figure 4-77. CURRENT-DRAIN SENSOR—Current drawn by dome light when door is open or by ignition when turned on triggers current-sensing stages Q1 and Q2 to start 555 timer and apply power to horn relay. Initial 15-s delay in sounding horn allows owner to enter car and open hidden switch S1 to deactivate alarm. If S1 is not opened during delay interval, horn sounds for about 90 s, then circuit automatically resets itself. C5 and R5 control duration of initial 15-s delay. C2 and R4 control total time that horn sounds.—R. F. Graf and G. J. Whalen, "The Build-it Book of Safety Electronics," Howard W. Sams, Indianapolis, IN, 1976, p 57-62.

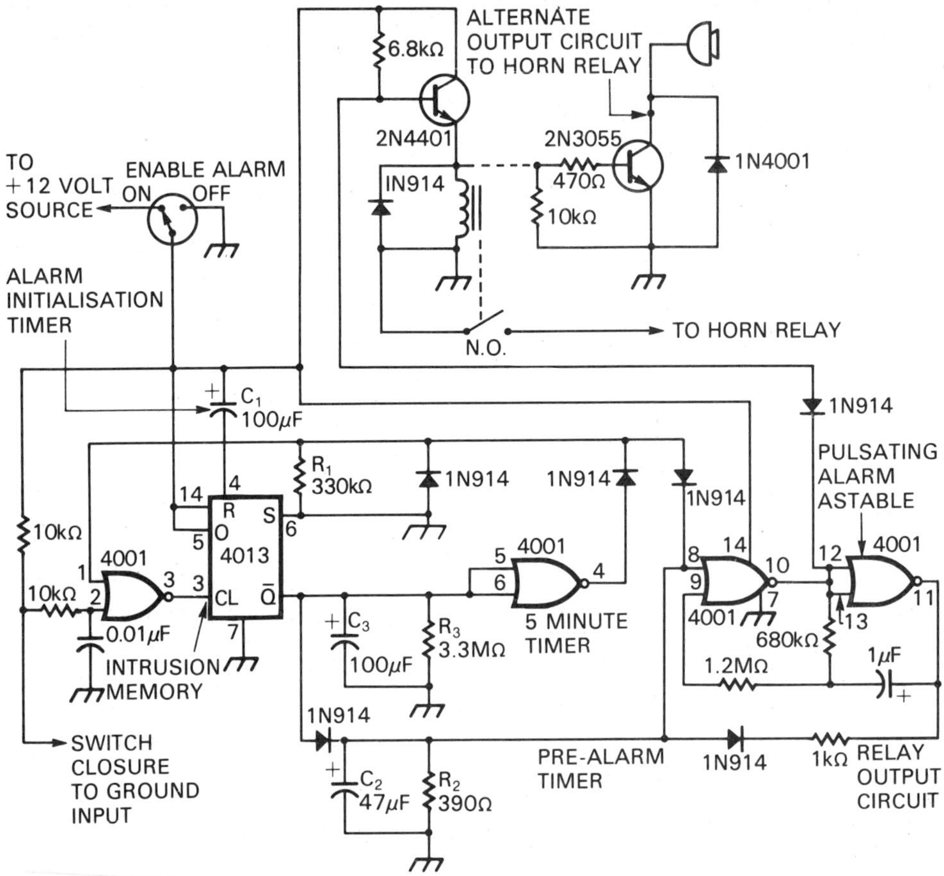

Figure 4-78. PULSED-HORN ALARM—Two CMOS packages incorporate multiple time delays to improve convenience and effectiveness of auto intrusion alarm. R1C1 gives 30-s delay for arming alarm after it is turned on by switch concealed inside car, to let driver get out of car. R2C2 gives 15-s delay before alarm sounds after door is opened, to allow driver to get back in car again and disable alarm. R3C3 turns off alarm in 300 S and resets alarm system for next intrusion. Car horn is pulsed 60 times per minute, so alarm would not be confused with stuck horn. Article tells how circuit works and gives detailed instructions for installation and connection to door and trunk switches.—G. Hinkle, Give the Hamburglar Heart Failure, *73 Magazine*, Feb. 1977, p 36-37.

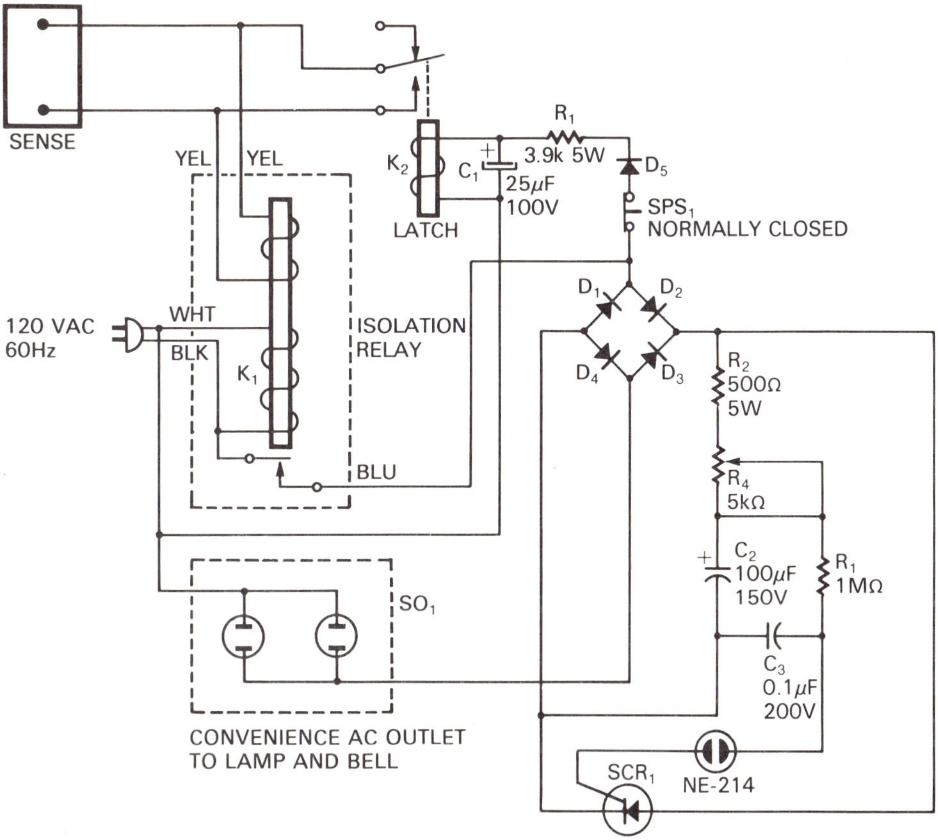

Figure 4-79. OPEN-CIRCUIT ALARM—Closing of door or window switch sensor or closing of normally open panic-button switch at bedside and other strategic locations in home trips alarm that sounds loud bell and flashes bright light on and off. Sensor shorts control windling of K1, allowing K1 to drop out and apply line voltage to alarm circuit. One AC path is through D5 which rectifies AC for energizing DC latch relay K2 to short sensor lines even though initiating sensor has opened. Simultaneously, AC is applied to diode bridge having SCR between DC legs. C2 starts charging through R2 and R4, and C3 charges through R3. When voltage across C3 reaches about 90 VDC, it fires neon and C3 discharges into gate of SCR. Full line voltage is then applied to lamp and bell plugged into load outlets. When C2 drops below holding current, SCR turns off during next AC cycle and load goes off until neon fires again. Setting of 5K pot R4 gives range of 15-80 flashes and horn pulses per second. To stop alarm, open SPST switch momentarily.—R. F. Graf and G. J. Whalen, "The Build-it Book of Safety Electronics," Howard W. Sams, Indianapolis, IN, 1976, p 75-80.

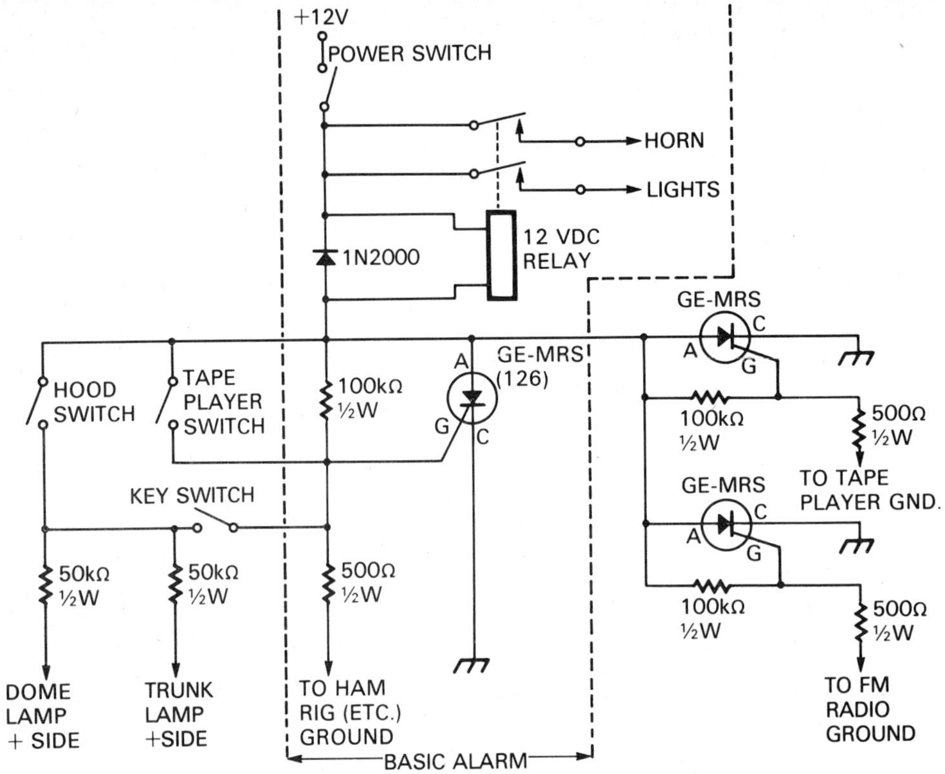

Figure 4-80. WIRE-CUTTING ALARM—SCR normally acts as open circuit in series with 12-VDC alarm relay because grid is made negative by voltage divider consisting of 100K in series with 500 ohms. If ground on 500-ohm resistor is removed, as by removal of tape player or CB set from car by thief, gate becomes more positive and SCR conducts, to energize relay, sound horn, and make headlights shine brightly. Additional triggering SCRs or alarm switches can be added as shown outside of dashed area for basic alarm.—A. Szablak, Another Burglar Alarm, *73 Magazine*, May 1974, p 45-46.

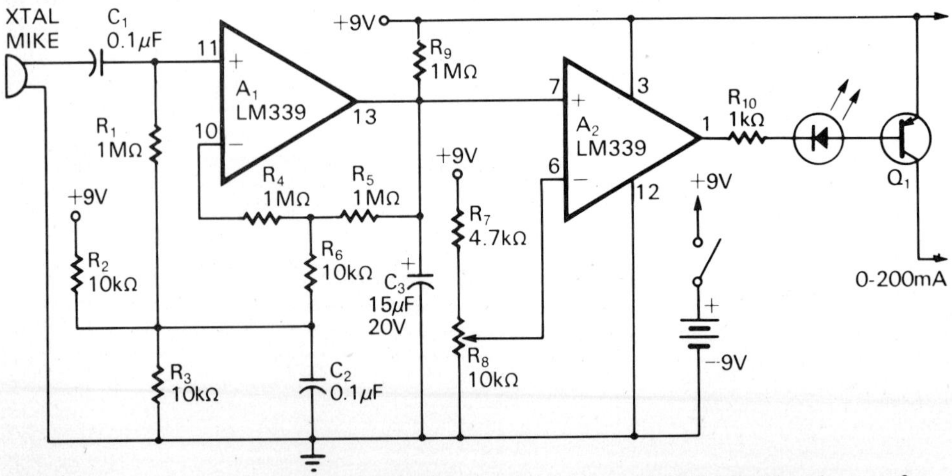

Figure 4-81. SOUND-ACTIVATED SWITCH—Can be used as sensor for burglar alarm or for turning on surveillance tape recorder to monitor conversations. R_8 is adjusted to give desired sensitivity at which A_2 triggers switch Q_1 to provide 200-mA load current and turn on indicator LED. First section of LM339 quad comparator serves as amplifier and detector providing gain of 100. Second comparator compares DC output of first with reference level selected by R_8.—D.R. Morgan, Sound Turns Switch On, *EDN Magazine*, Aug. 5, 1978, p 82 and 84.

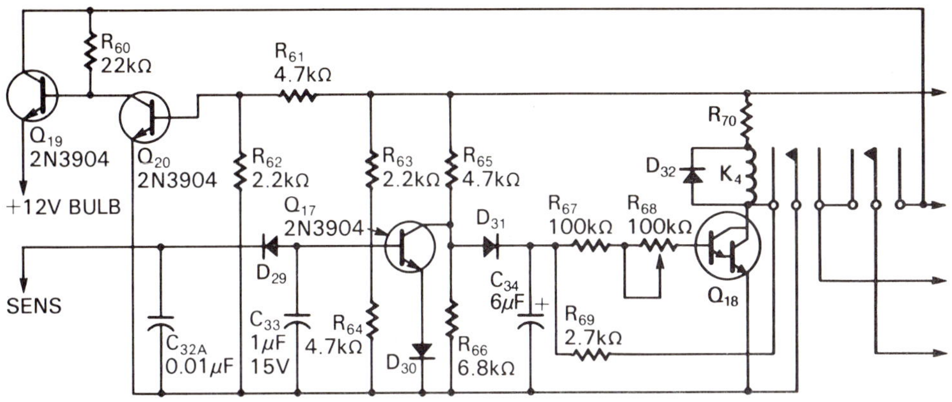

Figure 4-82. ALARM DRIVES PAGING BEEPER—Complete protection of vehicle is provided by multiplicity of door-switch, mat-switch, vibration, motion, and other sensors connected to common sensor input of alarm switching circuit that controls radio pager, 1-W GE Voice Command II transmitter operating around 147 MHz, 100-W electronic siren, and power horns. Closing of contacts in any sensor grounds common input (assuming keylock switch has been closed to arm circuit by applying +12 V), applying power to siren and pager system. Range is about 1 mi for Motorola Pageboy II cigarette-pack-size pager receiver. Article describes construction, operation, and installation in detail and gives complete circuit of pager.—J. Crawford, Build a Beeper Alarm, *73 Magazine*, Oct. 1977, p 68-77.

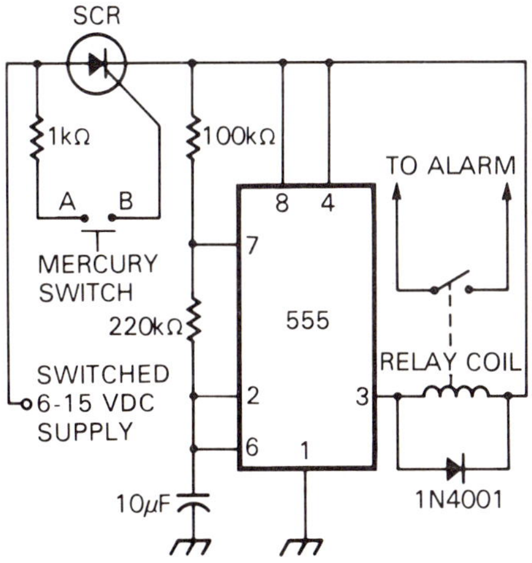

Figure 4-83. BEEPER—Intermittent alarm using 555 timer can be set to energize horn, lights, or other signalling device at any desired interval when tripped. When used on auto, sound cannot be mistaken for stuck horn. Choose SCR rating to handle current drawn by relay and timer. If alarm draws less than 200 mA, relay is not needed.—W. Pinner, Alarm! Alarm! Alarm!, *73 Magazine*, Feb. 1976, p 138-139.

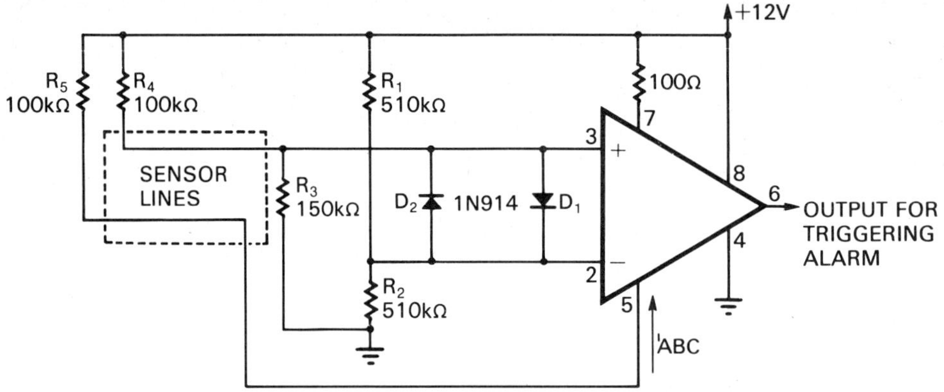

Figure 4-84. OPEN/SHORT/GROUND ALARM—Pin 6 of CA3094 IC is high for no-alarm condition. When any one sensor line is open, is shorted to other line, or is shorted to ground, output of IC goes low and resulting output current serves for activating alarm system.—E. M. Noll, "Linear IC Principles, Experiments, and Projects," Howard W. Sams, Indianapolis, IN, 1974, p 316-317.

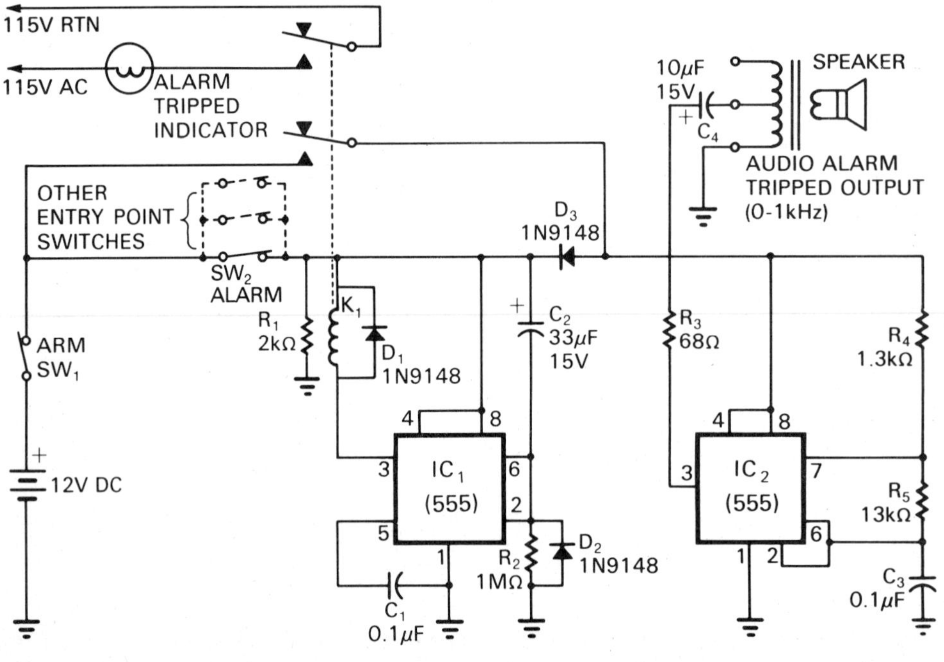

Figure 4-85. LATCH-ON ALARM—Alarm cannot be shut off for 12 s, with delay provided by IC, connected as mono, even if trip condition is immediately removed. Developed for home, garage, and auto burglar alarm systems having any number of trip switches. Circuit is self-resetting after delay interval, has high resistance to false alarms other than direct entry, and operates even when AC power fails. Circuit is latched through contracts of K_1. On standby, battery drain is zero.—J. D. Long, Burglar Alarm is Effective, yet Simple and Inexpensive, *EDN Magazine*, Dec. 20, 1974, p 50-51.

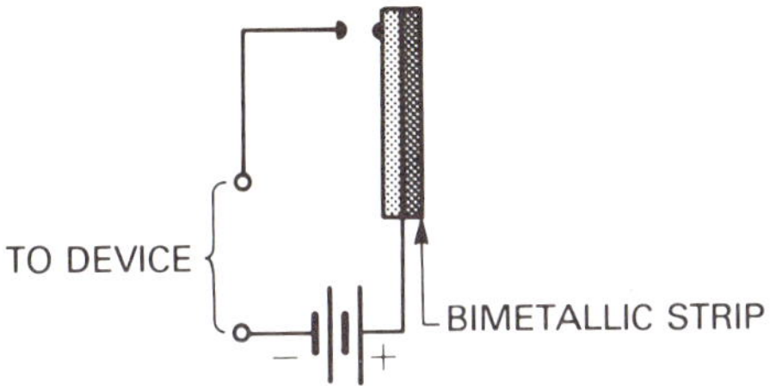

Figure 4-86.

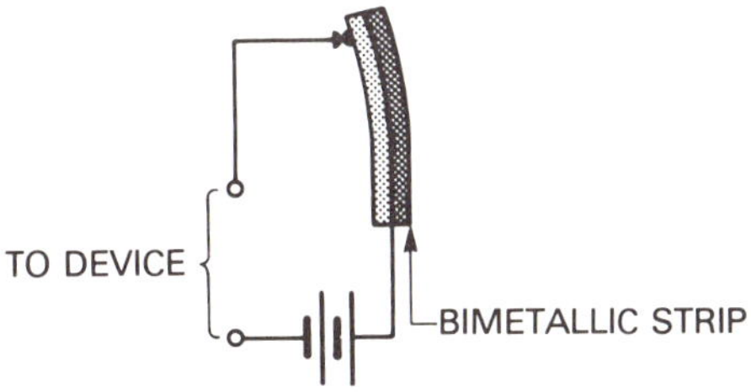

Figure 4-87.

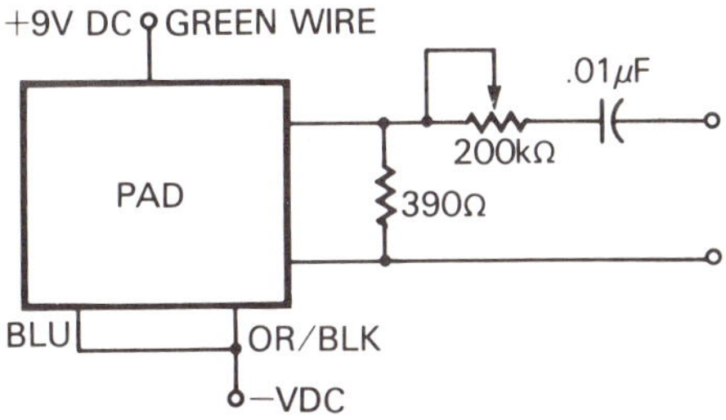

Figure 4-88.

Figure 4-89.

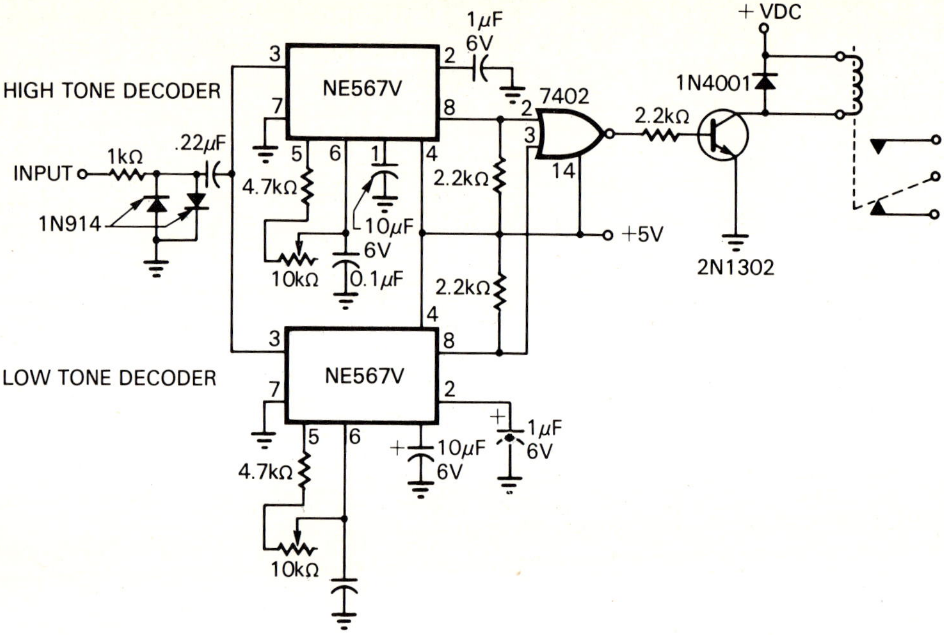

Figure 4-90.

LOW TONES	HIGH TONES		
	1209Hz	1336Hz	1477Hz
697Hz	1	2	3
770Hz	4	5	6
852Hz	7	8	9
941Hz	*	0	#

Figure 4-91.

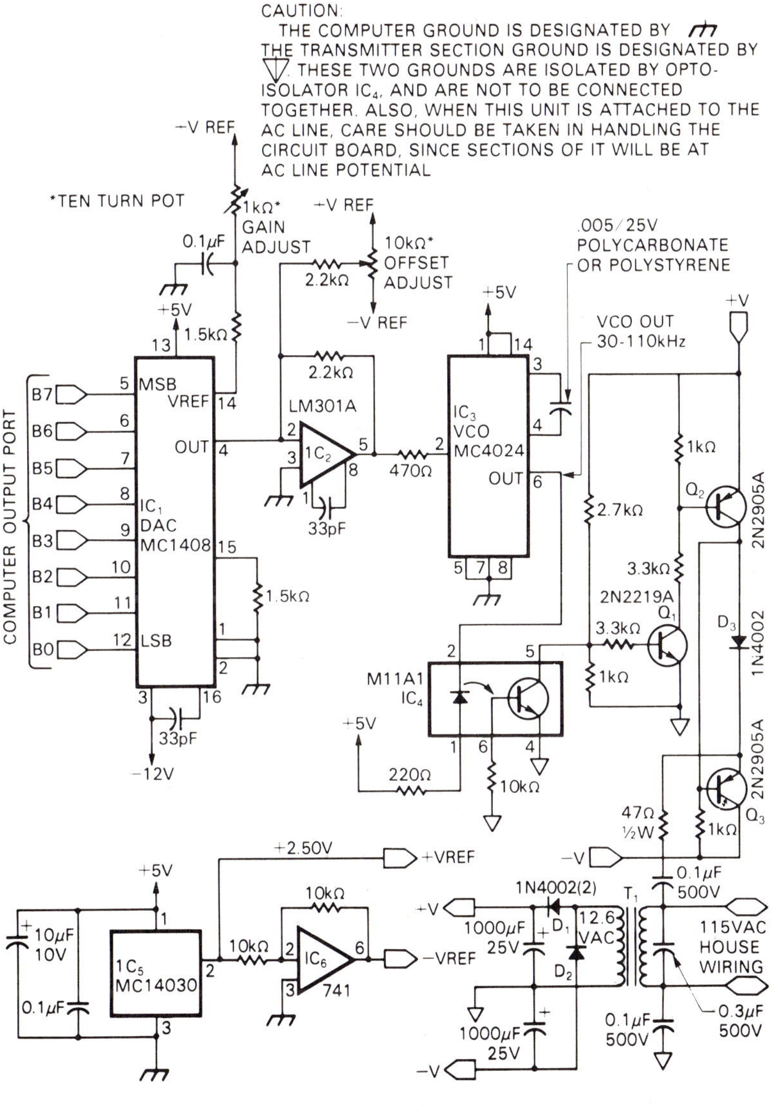

Figure 4-92. CARRIER-CURRENT TRANSMITTER—Modulates existing house wiring with high-frequency signals that can be detected by special receivers plugged into any AC outlet, for control of appliances by home computer. Applications include turning house lights on and off during owner's absence on elaborate time schedule programmed into computer. IC1 converts 8-bit data word from computer to proportional analog output current. This is converted to voltage by IC2 for control of VCO IC3 that gives frequency proportional to voltage. With values shown, range is about 30 to 110 kHz, with 256 discrete increments of frequency. Thus, input code 00000000 gives 30 kHz, 00000001 gives 30.3 kHz, and 01000000 (decimal 64) gives 49.2 kHz. Signal is applied to house wiring by 0.5-W power amplifier Q1-Q3, using optical coupling through IC4 to prevent computer circuit from interacting with house wiring. Supply voltage ±V is 11 to 13 V. T1 is 12.6-VAC 300-mA filament transformer. IC5 is 2 to 2.5 V reference ship such as MC1403U. System uses one frequency to turn receiver on and frequency 4 kHz above or below in 8-kHz band to turn it off, for maximum of ten control channels in system. Article covers calibration of transmitter.—S. Clarcia, Tune in and Turn on, Part 1: A Computerized Wireless AC Control System, *BYTE*, April 1978, p 114-116, 118, 120, and 122-125 (Part 2—May 1978, p 97-100 and 102.

TRANSMITTER

C_1,C_2 — 0.02μF DISC CAPACITOR
C_3 — 0.05μF DISC CAPACITOR
C_4 — 20μF @ 150V ELECTROLYTIC CAPACITOR
R_1 — 12kΩ, ½W RESISTOR
R_2 — 8.2kΩ, ½W RESISTOR
R_3 — 750Ω, 5W RESISTOR
R_4 — 400Ω, 5W RESISTOR
L_1 — 750μH CHOKE
L_2 — 420μH CHOKE
L_3 — 200μH CHOKE
D — DIODE 1N547 OR EQ.
Q — MOTOROLA HEP S3020 OR EQ.
S_1, S_3 — SPOT LARGE PUSHBUTTON (N.C. NOT USED)

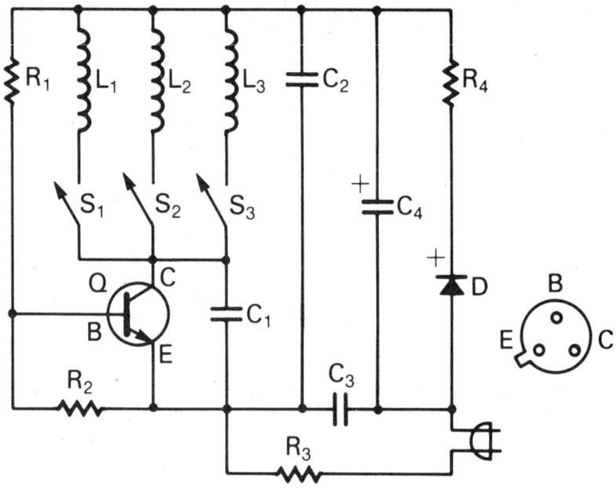

RECEIVER

C_1 — 0.0025μF DISC CAPACITOR
C_2 — 20μF @ 16V ELECTROLYTIC CAPACITOR
C_X — SEE TEXT
R_1 — 560Ω, ½W RESISTOR
R_2 — 1800Ω, 2W RESISTOR
R_3 — PHOTORESISTOR (CLAIREX CL 604L)
L — 508–816μH CHOKE
D_1, D_2 — 1N457 OR EQ.
Q — 2N3439 OR EQ.
K — 12VDC LATCHING RELAY (CORNL. DUBIR. 662.12)
NE — NEON NE-2

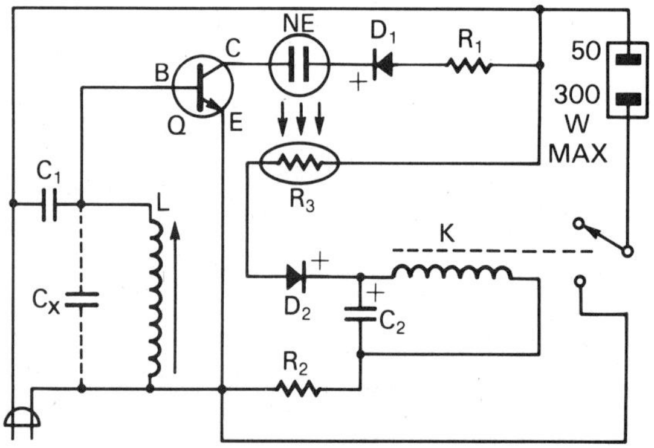

Figure 4-93. WIRELESS CONTROL—Choice of three lamps or appliances anywhere in house and garage, or even in neighboring home if on the same power transformer, can be turned on or off individually with three-channel transmitter that plugs into any wall outlet. Transmitter injects one of three tones (depending on button pushed) into house wiring. Receivers at locations of controlled devices are each tuned to one of carrier tones. Correct tone for receiver energizes neon lamp, and resulting light is picked up by photoresistor that energizes latching relay K for turning on controlled device. Relay is released by sending same tone again. Values of CX can be 0.005, 0.01, and 0.02 μF. Adjust slug of L2 for each receiver so neon comes on when assigned tone for that receiver arrives.—W. J. Hawkins, Three-Channel Wireless Switch—Use It Anywhere, *Popular Science*, Sept. 1973, p 96-99 and 121.

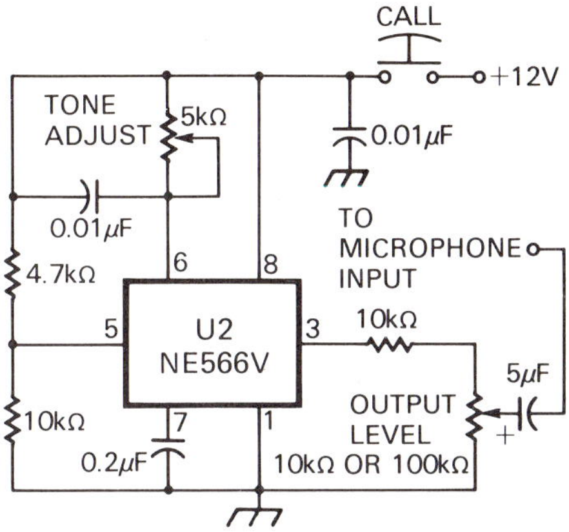

Figure 4-94. CONTROL-TONE GENERATOR—Uses NE566V PLL as tone generator, directly connected to microphone input of transmitter for activating loudspeaker in receiver being called. Operation is similar to that of paging units using selective call tones. Output level of PLL is adjusted so tone is same amplitude as voice.—K. Wyatt, Private Call System for VHF FM, *Ham Radio*, Sept. 1977, p 62-64.

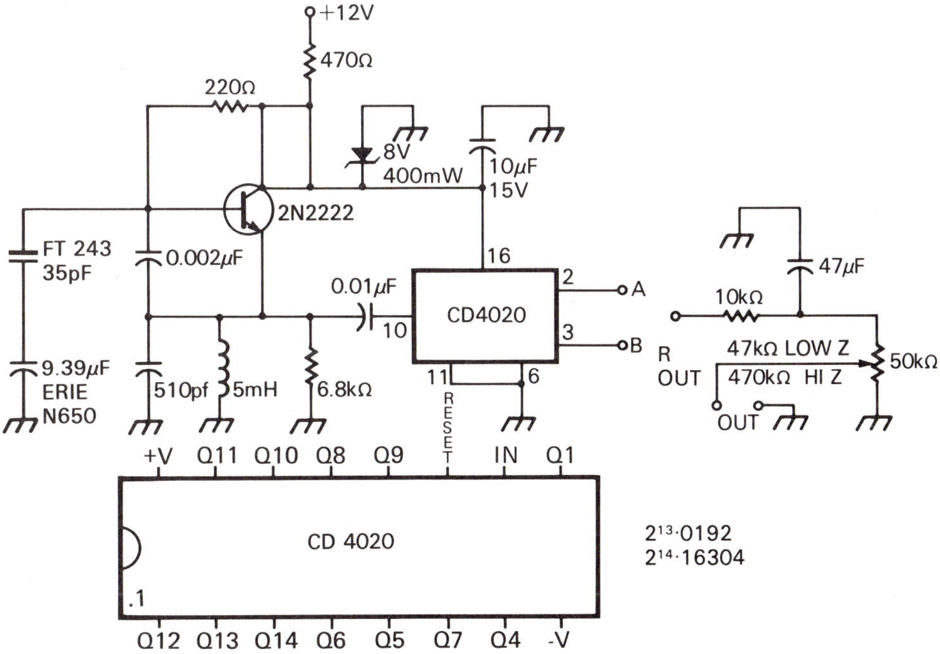

Figure 4-95. SUBAUDIBLE TONE ENCODER—Simple crystal-controlled oscillator drives CD4020 CMOS divider to give output frequency below normal voice range, for providing tone access to repeater. With 1.120-MHz crystal and division ratio of 8192, output is 136.7 Hz. Output circuit is RC filter that converts square wave to triangle, with pot setting level. Article tells how to choose crystals for other output frequencies and division ratios.—C. Haines, Jr., Go Tone for Ten, *73 Magazine*, Dec. 1976, p 22-23.

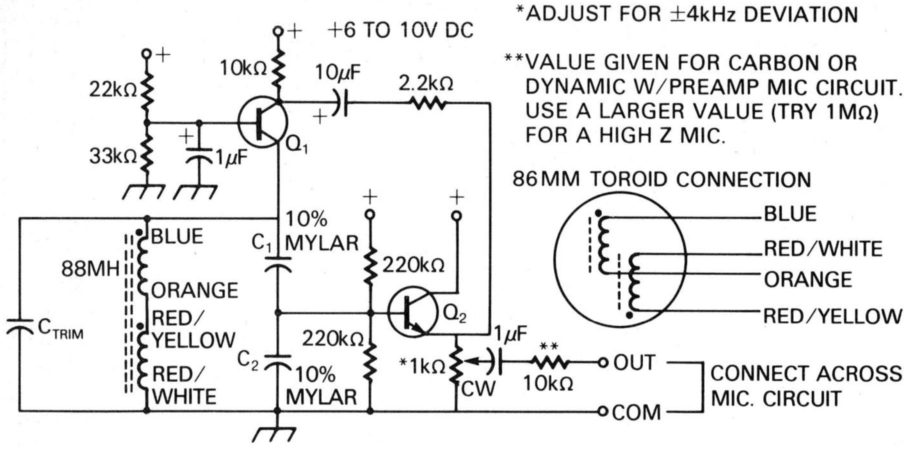

Figure 4-96. 1800-Hz COMMAND OSCILLATOR—Connects across microphone leads of FM transceiver, to produce control (command) tone for entry to repeater or for other purposes. Switch may be placed in supply lead if desired. Q1 can be 2N404 or one of transistors in 2N1303, 2N2904, 2N3638, 2N6516, or 2N6533 series. Q2 can be 2N1308, 2N2712, 2N3565, 2N3589, 2N6513, or equivalent. C_{TRIM} is 0.0062 μF.—Circuits, *73 Magazine*, April 1973, p 132.

5

APPLIANCES

INTRODUCTION

Heating, cooking and cooling appliances represent logical and readily available incendiary devices in almost every structure. In one package there is heat, fuel, and the potential for timed ignition.

Properly prepared, these devices will present the appearance of an accidental or mechanical malfunction fire cause. 'Properly prepared' includes not only careful preparation of the device itself, but also thorough analysis of the type, size, and general construction features of the subject structure versus the method or device to be employed as the ignition source.

Although most of these methods have a reduced pre-ignition potential when compared to some other methods described, the possibility of pre-ignition is always present and must be considered.

Natural Gas

Natural gas has several innate characteristics that make it interesting from both an investigator's viewpoint, as well as that of an arsonist. These characteristics include the fact that natural gas is lighter than air, has a lower flammable limit of 3.8% to 6.5%, and an upper flammable limit of 13% to 17% (air being 1). Gasoline has a lower limit of 1.4% and an upper of 7.6%, as a point of comparison. Unlike gasoline, however, natural gas has a #4 (highest) flammability rating, while gasoline only carries a #3 rating for flammability.

With these facts and characteristics in mind, one can readily determine that natural gas is very explosive, highly flammable, and will rise to the top of a room and fill "down" toward the floor.

Depending on the size of the house, there are several approaches that could be taken. If the structure is single-story with a crawl space, placing the heating appliances in the general living area, the "set-up" is simplified. If the structure is multi-story with a basement, and the heating equipment is in the basement, the possibility exists of simply closing off all doors to the first floor, as well as closing all basement windows, which will confine the "build-down" to the basement area. (This also presumes that the basement windows are tight.)

First the pilot light of the water heater must be lit. If the set-up involves a one-story house with a natural gas stove on the main level, in addition to the furnace and water heater, the pilot lights of both the stove and water heater remain lit. The water heater safety-cover plate is removed. At this point one of two things could be done, depending on the size and type of structure in which the set-up is taking place.

One method involves cutting or disconnecting the pilot light gas supply line (usually about ¼ inch diameter copper tubing). The valve controlling the pilot feed line is shut off and then the tube is cut or disconnected. Whichever method is chosen, the control valve is then turned on, allowing gas to escape into the surrounding environment, where it will fill "down" and be ignited by the pilot light of the stove and/or water heater.

The second method is to turn off the main valve to the furnace and loosen or disconnect the main supply line pipe (usually 2 inches diameter). The main valve to the furnace is turned on again and the leaking/fill-down process occurs. This latter method is much faster. (For this latter method the disconnect is made prior to the safety valve assembly.)

It will also be necessary to block any flue draft potential. This is readily accomplished by the use of a piece of cardboard or ¼ inch plywood wedged against the bottom of the flue. One of the problems with this "set-up" is that the odor of natural gas may be detected prior to ignition.

From the analyst's point of view, several things should be immediately obvious. If the ignition occurs in the lower explosive limits (L.E.L.) of the natural gas, a "low-order" explosion most probably will occur. The results will look like someone pushed out on the walls, up from under the roof (allowing it to fall back down in place), and most likely the window frames will blow out with glass intact. The odds are there will not be any fire.

At the other extreme, the possibility exists that the ignition will occur in the upper explosive limits (U.E.L.), which will result in blowing out the walls and roof, and a very intense fire will probably engulf the structure within a very short time.

Debris around the foundation walls may range from items located within a few feet (low order explosion) to items up to 100 feet or more away in a high order explosion.

There may also be reports of two explosions within the structure. This could be the result of a low-order explosion followed by a high-order explosion.

A thorough check of all pipe lines and interior supply lines must be made. If the threads are not stripped or if there are obvious indications of cutting tool marks on pilot light supply lines, the probability exists that the explosion/fire was induced, not accidental.

Flex Tubing and Natural Gas

In this "set-up" the water heater is fed by a rigid pipe mounted parallel to the floor joists in pipe hangers. A vertical pipe runs parallel to the water heater from the top of the appliance to the control valve at the base. These two pipes are connected by a piece of flex tubing. The vertical pipe is cut and the flex tubing is installed between the two "old" sections. See Figure 5-1.

Note that this could be done at any point where a horizontal and vertical pipe meet. (The reason for the presence of the flex tubing would most probably

Figure 5-1.

be given as a short piece of piping being bad and the easiest "handyman plumbing" to fit in was the piece of flex tubing.)

Bending (flexing) the flex tubing several times could create a crack, or a crack could be "induced" by a small hacksaw followed by flexing.

As soon as the hole is "induced," the escaping gas is lit with a match. Tests under normal pressure from gas lines inside a house indicate the flame will be approximately nine inches long when escaping from a "crack" six cm (approximately ¼ inch) long. If the flex tubing is placed against or very close to the floor joists or subflooring, it is obvious that ignition of the wood combustible will ensue. A presoak of the area immediately over the ignition area would also be a possibility. Presoaks of this nature are usually accomplished by coating the underside of the subflooring with lard or similar grease, and then spraying the grease with a flammable liquid.

Evidence for the analyst is somewhat limited. The most obvious evidence is the flex tubing itself; however, the presence of this tubing is not, in itself, conclusive. Examine the entire length of tubing. A close-up photograph of the "crack" is shown in Figure 5-2. Note that the "crack" is only about six cm long and there is evidence of a carbon build-up near the "crack," indicating that a fuel was burning near this area. As is obvious in the close-up photograph, a small nick by a hacksaw blade, followed by flexing to extend and expand the crack, leaves little evidence of a deliberately induced crack.

It remains, however, to be explained exactly how a crack in this area would accidentally ignite! The "fill-down" process would take many hours to occur from a leak this size, eliminating accidental ignition from nearby pilot lights.

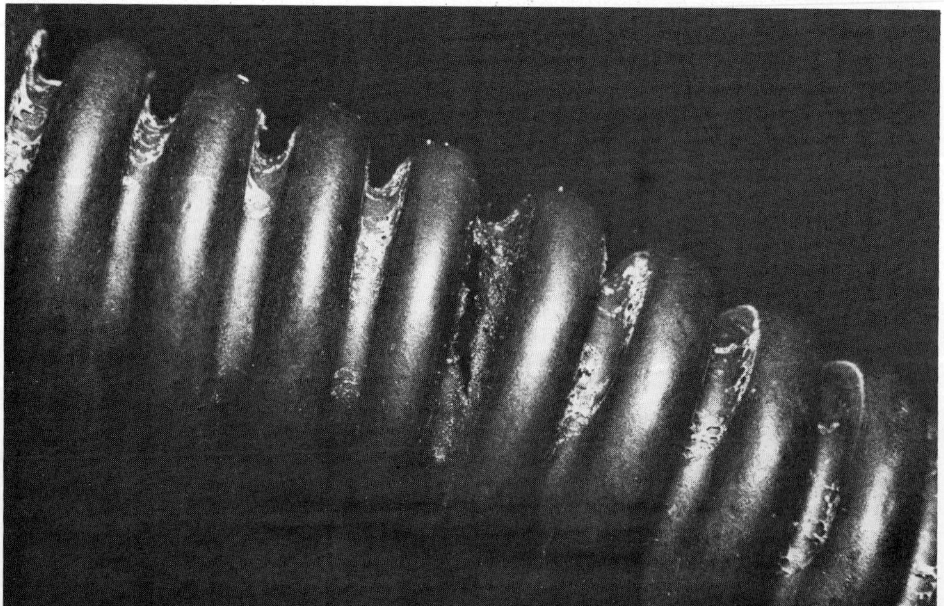

Figure 5-2.

Ovens

This one is quite simple. One places a tightly sealed one-gallon can of lacquer thinner, turpentine, or gasoline inside an oven (See Figure 5-3).

Leaving the oven door partially open, the heat is turned up to maximum. The liquid inside the can expands due to heat, the can explodes, and the liquid ignites from the heating elements inside the oven. The partially open door causes the oven to heat above the temperature it would normally reach.

Some sources have indicated the possibility of using paint; however, the most reliable liquid seems to be a mixture of approximately 40% gasoline and 60% turpentine or fuel oil.

From an analyst's viewpoint, the possibility exists that flammable liquid may remain in and around the oven, and/or the fire will not spread to the nearby cabinets or countertops. Examination of the coils and burner units, in addition to safety controls and valves, should reveal no malfunction and possibily no damage, except for exterior heat damage.

Figure 5-3.

Oil Pumps

In this method the first step is to turn the thermostat down to its lowest setting, after which the ignition transformer is removed from the blower/burner unit of the oil fired furnace. The air bleed/pressure gauge connection is then removed and the furnace is turned on by turning up the thermostat.

Oil will be pumped out of the air bleed tube and flood the floor. After sufficient oil has been pumped out onto the floor, the thermostat is again turned

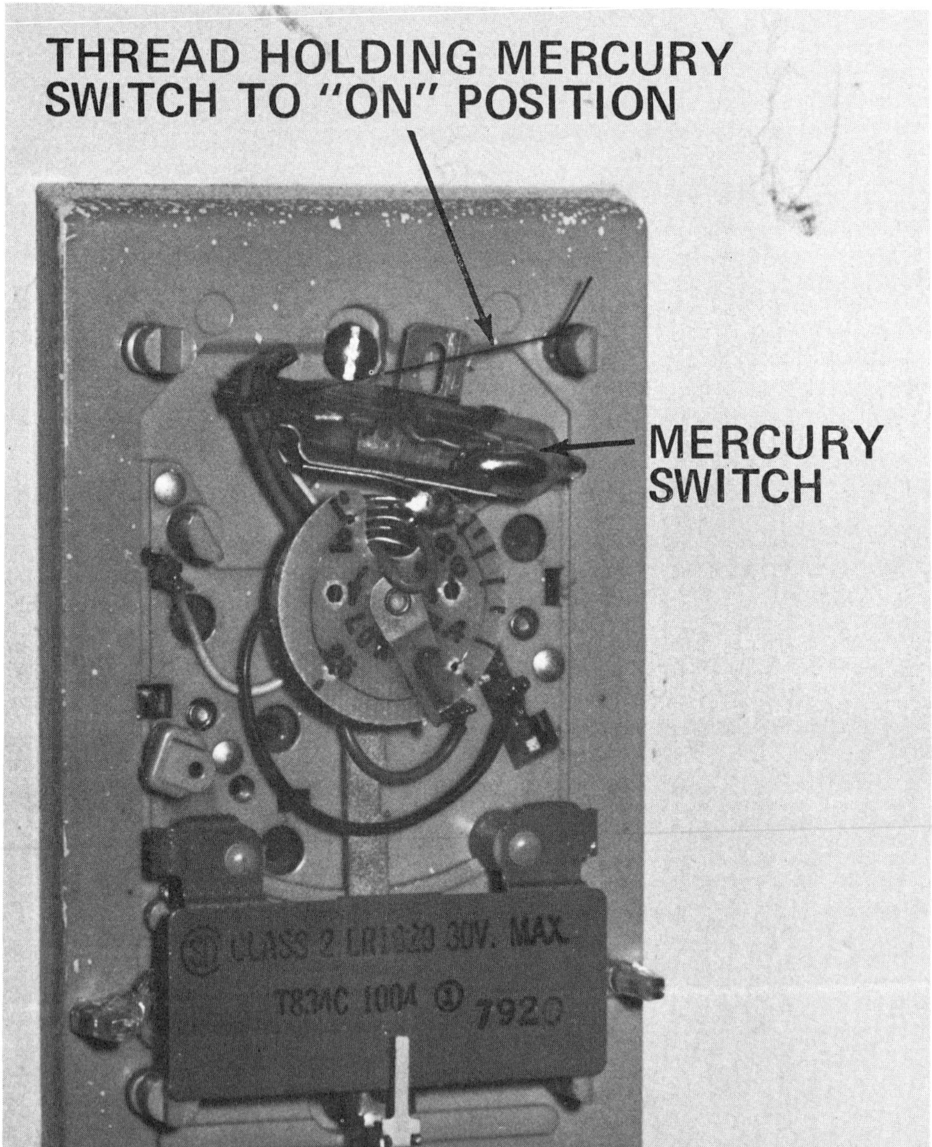

Figure 5-4.

down and the air bleed plug and ignition transformer are replaced. The oil on the floor is then ignited (by an open flame device) and the only evidence remaining might be some excess oil, which will probably be explained as a leak.

Oil Fired Hot Air Furnace I

A cotton or silk thread is prepared with a loop on one end that will hook over the mercury bulb of a thermostat (See Figure 5-4).

Figure 5-5.

The upper limit control is then by-passed with a jumper cable or by "hard-wiring" (which would appear more "normal" to one unfamiliar with the wiring of these controls). Figure 5-5 shows an upper limit control by-passed, which, in effect, takes it out of the circuitry, electrically.

After the upper limit control is by-passed and the thermostat is turned up (with the loop being positioned as indicated), the furnace will simply continue to run, creating an excessive amount of heat. It will take a few hours (time delay built in!); however, the furnace will eventually get red hot. If there are enough combustibles nearby (naturally or by design), they should ignite. The running fan will literally fan the fire to help spread it.

Nearby combustibles could be the wood structure, flammables atomized on the bottom of the subfloor or in plastic containers on top of the furnace, or fuel drained by the method described previously for oil pumps.

From an analysis viewpoint, the cotton or silk thread will disappear if the fire extends to the thermostat area. However, the thermostat will be turned up all the way, unless the person setting up the device realizes that tipping the mercury switch is all that is required. This means they would have only set the thermostat to a normal (68°, 70°, or 72° Fahrenheit) setting.

Once the make and model of the furnace are determined, the type of controls will then need to be determined. If this can be done, or if there is

Figure 5-6.

sufficient evidence to provide this information inside the controls themselves, the wiring can then be traced to determine the presence of alien wiring.

If the required information is unavailable, possibly a local repair service could identify the type of control and provide documentation on the proper wiring arrangements.

Electric Charcoal Starter

The electric charcoal starter represents a somewhat different way of starting a fire that could readily be blamed on carelessness or children, or both. The coil portion of the starter is placed in or on the combustibles to be ignited. The unit is plugged into the wall receptacle. Time, temperature, and fuel do the rest.

For the analyst, the presence of the electric starter in an unusual place needs to be explained. If the unit were inserted into stuffed furniture on a storage porch, an obvious answer would probably be children playing in the area.

Refrigerators

In this set-up the refrigerator used must have a coil/relay arrangement. Many newer refrigerators have a relay; however, it does not spark as much as older relays using coils and multiple contact points.

To prepare this device, the refrigerator is first cycled, that is, one "times" the cycles for "on" compressor and "off" compressor, along with the amount of time between off and on. Once this is known (by timing at least three cycles), the protective cover around the solenoid/relay is removed.

During the "off" cycle, the area underneath the refrigerator is soaked with a flammable liquid (usually 60% gasoline and 40% fuel oil). When the refrigerator cycles the spark from the making/breaking of the contact points ignites the flammable vapors in the area. Figure 5-6 shows a close-up of the relay coil with the cover removed.

Sometimes the set-up may also include setting a can of the flammable liquid mixture on top of the refrigerator. The bottom of the can will have a small hole punched through it so the flammable will continue to drip or run down the rear of the refrigerator cabinet. Most often this run pattern will follow the a.c. line cord down to the area of the relay coil. Ignition of the flammable vapors underneath the cabinet will result in vertical flame spread to the fuel source at the top of the cabinet, with subsequent spread to adjacent cabinets, kitchen walls, and ceiling.

Analyzing the fire debris will first provide a visual finding of no cover over the solenoid. There may also be evidence of fire on the lower inside of the refrigerator.

There may also be a "V" (possibly inverted) up the rear of the refrigerator (it could also have been plastic). There will be an obvious intense heat pattern in a rather localized area, with the possibility of metal sagging and distortion of the top. The presence of flammable liquid residue is also possible.

Electric Hot Plates/Heaters

This ignition device is nothing more than an a.c.-powered hot plate or electric (floor-standing type) room heater placed too close to combustibles. For best results, the heater is placed face down against furniture or carpets. (Newer carpets may not have the necessary flamespread to guarantee success.)

The possibility exists that the floor and furniture could be presoaked with a flammable liquid, which would aid in spread of the fire. However, this does leave the possibility of residue that could be detected by a lab.

In using an electric hot plate, it simply becomes a matter of removing the bottom cover so heat will radiate downward, or of placing the unit so the heating coils are in direct contact with some type of combustible. (See Figure 5-7)

Obviously the fire analyst will have difficulty proving this method to have been incendiary unless motive and intent can be shown. The remaining evidence is not considered "alien" or "abnormal" in a house. The most important item to look for, in the case of the hot plate, is the missing bottom cover.

Oil Fired Furnace II

This method is similar to other furnace methods, except that in this instance a two-stage oil pump is involved. This means that there are two fuel lines utilized: one is the supply line and the other is the return line.

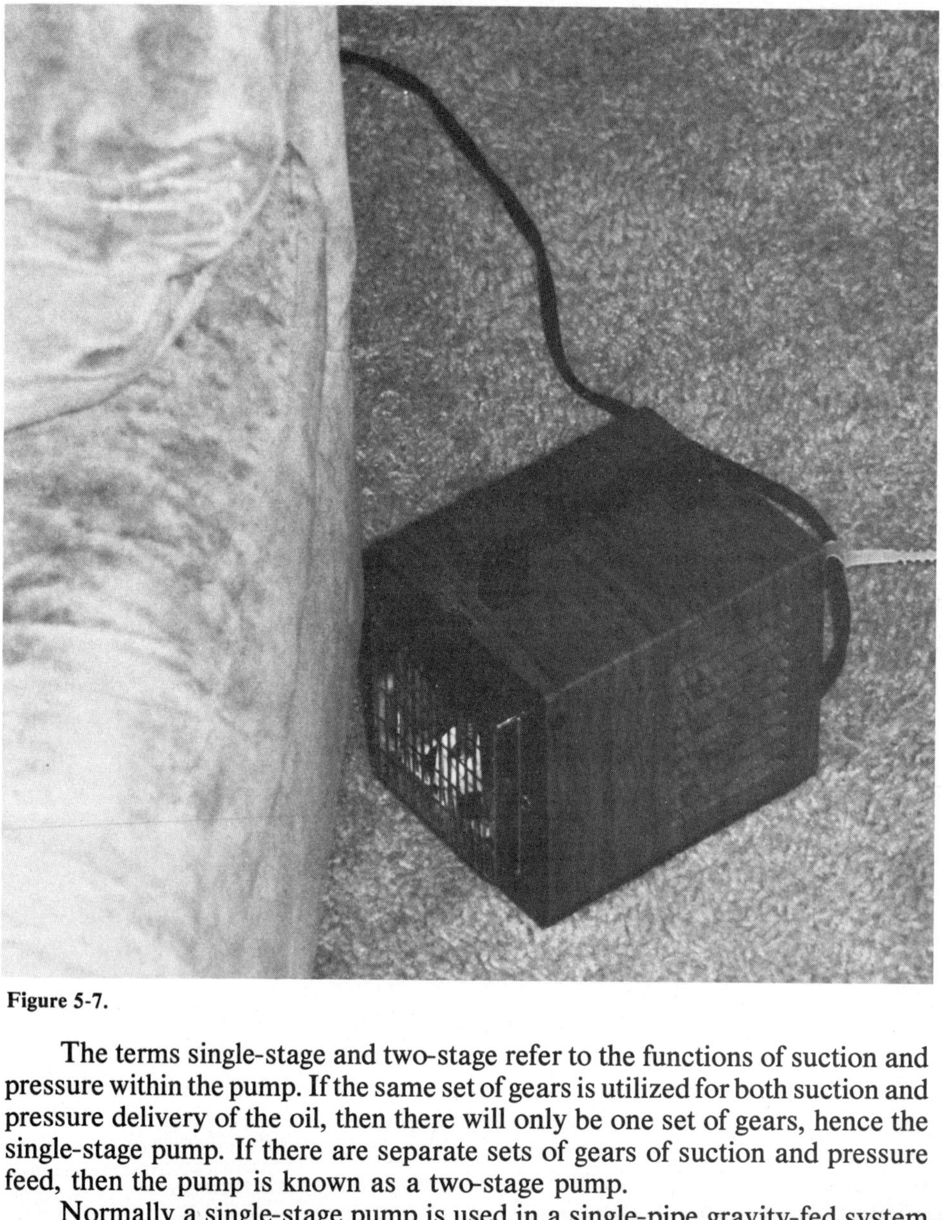

Figure 5-7.

The terms single-stage and two-stage refer to the functions of suction and pressure within the pump. If the same set of gears is utilized for both suction and pressure delivery of the oil, then there will only be one set of gears, hence the single-stage pump. If there are separate sets of gears of suction and pressure feed, then the pump is known as a two-stage pump.

Normally a single-stage pump is used in a single-pipe gravity-fed system and a two-stage pump is encountered where the supply tank is lower than the furnace/blower unit and it is necessary to pump the oil up (against gravity) from the tank.

In this system (two-stage), more oil is usually pumped up to the blower unit than is used and the return line serves to drain off the excess amount. For purposes of this device, the excess amount pumped up is of benefit in the set-up.

The thermostat is turned to its lowest setting. The ignition transformer is removed and the return line disconnected. With the ignition transformer still removed, the thermostat is turned up (demanding heat), which causes oil to be pumped out onto the surrounding floor. Simultaneously, some will be atomized into the fire chamber by the blower unit.

After two or three minutes of pumping, the thermostat is turned down, which stops the pumping and atomizing of fuel. The ignition transformer is replaced and the return line is reconnected, if desired.

Leaving the return line disconnected will cause oil to be pumped out onto the floor, which will "feed" the hostile fire after ignition. This would also be a strong piece of evidence when uncovered after the fire.

An alternative is to connect the return line loosely to the pump so it "appears" normal. This will still allow some fuel to the fire after ignition.

Ignition can be effected in several ways, the most usual being to simply demand heat by turning up the thermostat. This requires setting the thermostat at its highest, which could be evidence at a later point.

Exterior ignition of the fuel on the floor, coupled with thread on the thermostat and by-passing upper limit control, is an alternative ignition method that would be certain, and that would leave relatively little evidence.

From the analyst's viewpoint, there are several items which would prove that the fire was incendiary. On both the exterior and interior surfaces of the furnace cabinet, there would be signs of intense heat. The insulation blanket on the interior would most probably be destroyed. The helical inside the high limit control would probably be expanded to its outer limits, and be resting against the supporting structure.

Examination of the return line connection (assuming that this connection was loosely tightened) would provide evidence of a carbon build-up as a result of oxygen being available at the fuel/heating source, which allowed the fuel to burn inside the connection.

Obviously, the likelihood of all of these things happening in a single process is nil when one considers, for example, the heat necessary to: expand and distort the helix; permeate the metal cabinet to burn the insulation blanket (which doesn't burn during normal operation of the furnace); and to support combustion inside a connection from which the fuel oil normally didn't leak.

Oil Burner Controls

An older type of oil burner control can still be found in quite a few installations throughout the country. Although this method may not result in a hostile fire situation, it will definitely create a hazard and, if used in conjunction with other methods, will contribute greatly to the spread of fire and continuation of a heat source for the fire.

A small piece of wood or metal is used to block both relays simultaneously, which causes the furnace to overheat. In addition to the blockage of the relays, the safety switch is tied together (usually with string) to preclude "dropout," which results in no "safety-lock-out" so the unit can't shut down.

In lieu of string, the safety switch could be by-passed by wiring dressed to look "normal." Tracing the wires and wiring connections after the fact would confirm or deny that this was done.

From an analyst's point of view, the evidence would be virtually non-existent if the fire was successful. The wood or metal pieces will normally fall out from between the relay coil and frame mounts, which allows them to either be consumed or melted in the flames.

In effect, this procedure by-passes the limit controls and, from the time of wedging of the pieces and tying-off of the lock-out switch, the procedure could be the same as in Oil Fired Hot Air Furnace I.

6

MATERIALS FORMULAS

INTRODUCTION

This section is included as a reference to be used when a question arises as to the probability of certain materials or chemicals having contributed to the cause of the fire.

Example: You are called to the scene of a closet fire. Upon arrival you find that a clothes dryer has been placed in the closet and among the materials and debris you find fabric softener bearing a warning label: "Do not mix soap, detergent, or any type of bluing agent with this fabric softener." Scene processing reveals that soap, detergent, and a bluing agent were spilled on a pile of silk and cotton materials at the point of origin. Spontaneous combustion? Probably not.

Referring to fabric softeners, you find that it is a water-based product, which would appear to render it harmless as regards this particular fire cause. The warning most likely means that mixing these various chemicals would result in an adverse effect of the desired cleaning results, as opposed to causing a violent chemical reaction.

This list is not exhaustive and, in some instances, manufacturers will be using ingredients other than those listed. This is only a guide and each case encountered must be examined on its own merits. Not all of the ingredients listed under each heading will be found in each product, as combinations and percentages of ingredients vary from manufacturer to manufacturer.

ACETONE
Isopropyl alcohol distilled from fermentation of grain. Used in gums, resins, lacquers, thinners and ink removers. Methyl acetone is a mixture of about 50% acetone, 30% methyl acetate and 20% methanol, and is used in lacquers, cellulose plastics, paint removers, and for coagulating latex.

ACRYLIC RESINS
Esters of acrylic acid
Methacrylic acid
Ethyl acrylate and methyl acrylate
Used in adhesives, finished, laminated glass, and molded products.

ACRYLONITRILE
Also known as vinyl cyanide. Composed primarily of hydrocyanic acid and acetylene.

ACRYSTALYTE CLEAR FLOOR FINISH
Acrylic polymers
Resin
Polyethylene
Ammonia
Morpholine

Plasticizers
Water

AFTER SHAVE LOTION
Boric acid
Glycerine
Denatured alcohol
Water
Perfume
Benzophenone
Dyes

AJAX® CLEANSER
Trichloro-s-triazinetrione
Sodium dodecyl benzene sulfonate
Trisodium phosphate
Sodium bromide
Silica

ALUMINUM CLEANER
Powdered alum
Trisodium phosphate
Water

ALUMINUM AND STAINLESS STEEL CLEANER
Silica
Sulfamic acid
Sodium silicofluoride
Corn starch
Anionic surfactant
Perfume

ALUMINUM WELDING FLUX
Potassium chloride
Sodium chloride
Lithium chloride

ANALGESIC RUBBING LOTION
Methyl salicylate
Camphor
Menthol
Mineral oil
Denatured alcohol

ANIMAL REPELLENT
(For indoor furniture)
Methyl noyl ketone
Sodium nitrite

ANTIFREEZE
1. Methanol
Isopropyl alcohol
Rust inhibitors (Borates)
Alkali metal borates and phosphates
Dyes
2. Ethylene glycol
Other glycols
Rust inhibitor (borax)
Caustics
Sodium mercaptobenzothiazole
Dyes
3. Miscellaneous
Calcium chloride-sodium chloride and inhibited propylene glycol are used for refrigeration brines. Calcium chloride solutions are also used in tractor tires for added weight.
Glycerol and ethyl alcohol are used as antifreezes for cosmetics.
Lubricating oils have been used as cooling agents in specially prepared engines.

ANTIQUING AGENTS
1. Antiquing glaze
Magnesium carbonate
Silica
Silicates
Linseed oil modified polyurethane resin
Alkydized synthetic oil
Petroleum distillates
Driers
Tinting colors
2. Antiquing base coat
Titanium dioxide, type II
Phthalocyanine blue
Red iron oxide, class A
Calcium carbonate
Vinyl acrylic copolymer resin
Additives
Ethylene glycol
Phenyl mercuric acetate
Water

ANTIPERSPIRANT DEODORANT
Aluminum chlorohydrate
Isobutane
Propane
Cyclomethicone
Di-butyl phthalate
Isopropyl palmitrate
Quaternium-18 bentonite
Zinc oxide
Denatured alcohol
Water

ANTISEPTIC (*See* Disinfectants)

ASBESTOS
Actinolite (a mineral)
Chrysotile (a fibrous variety of serpentine

AUTOMOBILE ENGINE CLEANERS
1. Ethylene dichloride
Cresol (low boiling cresylic acids may also be used)
Oleic acid
Potassium (or sodium) hydrozide
Water
2. Methylene chloride
Perchlorethylene
Stoddard solvent
3. Perchlorethylene
Trichlorethane
Methylene chloride
Petroleum solvents
Chlor-aromatic solvents
Detergent
Emulsifier
May also contain:
Alkali
Corrosion inhibitors
Essential oils
Lubricating oils
Pine oil
Versene
4. Trisodium phosphate
Sodium bicarbonate
Sodium metasilicate

AUTOMOBILE FUEL ICE PREVENTATIVE
Denatured alcohol
Lightweight lubricating oil
Pine oil
Triethanolamine

AUTOMATIC TRANSMISSION FLUID
Mineral oils
Oxidation inhibitors and detergents
Zinc dialkyl dithiophosphate
Phosphorus pentasulfide-terpene and additional products
Barium and calcium petroleum sulfonates
Barium and calcium alkyl phenolates
Ashless detergents
Pour depressants and viscosity improvers
Polymethacrylate esters
Polyvisobutylenes
Hydrocarbon was-napthalene condensation products
Anti-wear agents
Organic borates
Antifoam agents
Polysiloxanes
Sealant
Triarylphosphate
May also contain dyes

AUTOMOBILE WINDSHIELD CLEANER
Calcium carbonate
Sodium bicarbonate
Diatomaceous earth
Water

BAGWORM SPRAY
Toxaphene
Xylene
Kelthane

BAKELITE® (*See* Formica)

BATH OIL
Isopropyl myristate
Mineral oil
Lanolin oil
Fragrance

BLEACH PACKETS
Alkaline mixture of inorganic salts:
Sulfate
Phosphate
Carbonate
Synthetic detergent

BOILER CLEANERS
Liquid
Hydrochloric acid
Phosphoric acid
Oxalic acid
Sulfamic acid
Citric acid
Ammonium or sodium fluoride
Powder
Asbestine
Sodium carbonate
Trisodium phosphate
Potassium dichromate
Quebracho

BORIC ACID
Hydrochloric or sulphuric acid
Borax

BRAKE FLUIDS
Lubricant
Castor oil
Castor oil soap
Butyl or glyceryl ether of polyoxyethylene propylene glycol
Polyprophylene glycol
Solvent
Methyl, ethyl, and butyl ethers of ethylene glycol and related glycols
Brake fluids may also contain:
Butylene glycol
Diethylene glycol
Ethylene glycol
Hexylene glycol
Polyethylene glycol
Inhibitors
Amine soaps
Potash soaps
Borax
Antioxidants
Bisphenol A
Hydroquinone
Dyes

BRASS, COPPER, AND CHROME CLEANER
Sodium bicarbonate
Sodium metasilicate
Trisodium phosphate
Ammonia

BUBBLE BATH
Sodium lauryl sulfate
Sodium sesquicarbonate
Alginate (sodium)
Perfume
Water
Alcohol
Phosphoric acid

CANVAS FIREPROOFING COMPOUND
Ammonium phosphate
Ammonium chloride
Water

CARBURETOR CLEANERS
1. Tall oil
Cresol
Potassium hydroxide
Ethylene dichloride
Sodium chromate
Ammonium oxalate
Alcohol
Water
2. Ethylene dichloride
Cresol
Oleic acid
Potassium hydroxide (sodium hydroxide)
Water
3. "Odorless" carbon removers
Aliphatic or aromatic hydrocarbon solvents with oil soluble wetting agents

CARPET CLEANER
Wholewheat flour
Mineral oil
Aluminum stearate
Salicylic acid
Water

CAT BOX LITTER
Aluminum-magnesium silicate
Iron oxide

CATALIN (*See* Formica)

CEMENT
1. Portland
Calcereous and argillaceous minerals heated. High calcium limestone and shale or rock
2. White
Pure calcite limestone, chalky limestones (France and England), bauxite, and alumina

CHARCOAL LIGHTERS
Naphtha
Perfume
Aromatics
Kerosene
Mineral spirits

CHEMI-LUME® LIGHTS
Fluorescein rhodamine B
Methyl cellulose

CHEMISTRY SETS
1. Ingredients graded as very low toxicity:
Bean seed
Calcite

Calcium carbonate
Calcium sulfate
Cochineal
Copper wire
Diatomaceous earth
Epoxidized soybean oil
Granulated zinc
Graphite
Gum arabic
Iron metal (powdered)
Logwood
Moscovite
Nichrome wire
Nickel steel wire
Phenolphthalein solution
Polyvinyl alcohol
Powdered charcoal
Powdered iron
Rhodamine B; triphenylmethane dye
Sodium sulphate
Starch
Sulphur
Thionine
Zinc metal

2. Ingredients of low toxicity
Acetic acid
Acetyl-tributyl citrate (plasticizer)
Ammonium chloride
Bronze powder
Calcium chloride
Calcium monophosphate
Chrome alum (chromium potassium sulfate)
Giberellic acid
Gypsum (calcium sulfate, hydrous)
Magnesium sulfate
Manganese sulfate
Mineral oil
Oregon balsam
Shellac
Sodium bicarbonate
Sodium bisulfite
Sodium chloride
Sodium ferrocyanide
Sodium iodide
Sodium thiosulfate
Strontium chloride

3. Potentially toxic ingredients:
Aluminum sulfate
Ammonium bydroxide
Azodicarbonamide
Azurite (basic copper carbonate)
Benedicts solution (alkaline copper sulfate)
Borates
Calcium hypochlorite
Calcium nitrate
Calcium oxide
Cedarwood oil redistilled
Cobalt chloride
Diglycolsterate
Ethanol
Ferric ammonium sulfate
Ferrous ammonium sulfate
Gold solution (possibly not in sets since 1962)
8-Hydroxyquinoline
Melamine formaldehyde
Nickel ammonium sulfate
Ninhydrin
Potassium chloride
Sodium carbonate
Sodium salicylate
Sodium tetraborate
Stilbene
Strontium tetraborate
Strontium nitrate
Tannic acid
Tartaric acid
Trisodium phosphate
Zinc sulfate

CHRISTMAS SNOW SPRAY

Waxes
Fatty acids
Perfume
Chlorinated hydorcarbon solvents
Petroleum distillates
Nitrocellulose or hydrocarbon resins
Freon or isobutane propellants

COFFEE POT CLEANERS

1. Tripolyphosphate
Sodium metasilicate
Alkyl aryl sodium sulfonate
Ethoxylated tridecyl alcohol
Sodium bicarbonate
2. Sodium phosphates
Sodium silicate
Sodium perborate
Sodium carbonate
Surfactant
3. Silicate of soda

"Santocel"
Sodium sesquicarbonate
Soda ash
Sodium perborate
CF-10
Sodium citrate

CONTACT CEMENT
1. Resin latex type
   Neoprene latex
   Resin gum emulsion
   Water
   Casein
   Traces of ammonium hydroxide, sodium-O-phenylphenate
2. Rubber cement
   Hexane
   Rubber
   Solvents, including toluene, ketone (acetone), or chloroethene
3. Solvent type
   Neoprene rubber
   Modified phenolic resin
   Magnesium oxide
   Zinc oxide
   Solvent blends, including toluene, hexane-type
      Aliphatic cut, methyl ethyl ketone and 1,1,1-trichloroethane for nonflammable adhesives

CRAYONS
1. Children's crayons
   A. Wax type:
      Wax
      Paraffin and other waxes
      Pigments
   B. Pressed type:
      Paraffin and other waxes
      Pigments and extenders
2. Industrial crayons
   Lead chromate
   Pigments

CREOSOTE PRESERVATIVES
Coal tar creosote oils
Coal tar acids

CRYSTALLIN (*See* Formica)

CUTTING OILS
1. Cutting oils—chemical coolants
   Water
   Rust inhibitors (sodium nitrate, ethanol amines, potassium or sodium soaps)
   Lubricity agents (polyether glycols, alkyl-phenol-ethylene oxide condensation products)
   Bactericides (chlorophenols, organic mercurials, iodine compounds, formaldehyde releasers, quaternary ammonium compounds)

NOTE: Chemical coolants are usually diluted with 50 to 150 parts water to 1 part coolant to form products used in metal-working operations.

2. Cutting oils—insoluble type
   Mineral oils, including sulfurized mineral oils
   Fatty oils, including sulfurized fatty oils
   Sulfur (combined and suspended)
   Chlorine (chlorinated paraffins, rarely chlorinated aromatics)
   Phosphorus (organic phosphates and phosphites)
3. Cutting oils—soluble type
   Mineral oils
   Water
   Emulsifiers (sodium and amine soaps, sodium sulfonates, napthenates, rosinates)
   Coupling agents (alcohol, glycol ethers, glycols)
   Rust inhibitors (amines, fatty oils, sulfurized fatty oils)
   Bactericides (chlorophenols, organic mercurials, iodine compounds, formaldehyde releasers, quaternary ammonium compounds)

NOTE: Soluble cutting oils are usually diluted with 10 to 75 parts water to 1 part oil to form emulsions used in metal-working operations.

DAIRY SPRAY
Petroleum distillates
Pine oil
Pyrethrins

DEGREASERS
Household liquid:
1. Combinations of anionic and nonionic synthetic surfactants, usually about 15% to 20% with remainder being water.

2. Water solutions of coconut fatty acids in combination with previously mentioned surfactants.
3. Petroleum distillates (which may contain methyl polysiloxane, tall oil fatty acids)

Powder (carbon tetrachloride, kaolin, naphtha)

Vapor (perchlorethylene, carbon tetrachloride, trichloroethylene)

DETERGENTS
Alcohol sulphonates
Sodium silicate
Sodium lauryl sulphate
Alkyl benzene sulphonates

DIATOMACEOUS EARTH
Granular minerals of opaline silica

DIP-IT®
Sodium perboate
Sodium carbonate
Sodium silicate

DISHWASHING DETERGENT (Automatic)
Soda ash
Sodium metasilicate
Trisodium phosphate

DISINFECTANT (LYSOL®)
Phynylphenol
Ethyl alcohol
CO-2
N-Ethyl morpholinium ethylsulfates
Creosote oil
Cresylic acid
Formaldehyde

DRAIN CLEANER
Sodium bicarbonate
Sodium chloride
Cream of tartar

DRY-CELL BATTERIES
1. Conventional "D" size cell battery
   Carbon
   Manganese dioxide
   Ammonium chloride
   Zinc chloride
   Cornstarch
   Zinc (container)
   Mercuric chloride
   Water
2. Alkaline Manganese "D" cell
   Zinc metal
   Mercury metal
   Potassium hydroxide
   Carboxymethyl cellulose
   Zinc oxide
   Manganese dioxide
   Carbon
3. Mercury cell battery 401 (hearing aids, pen lights, and small transistor radios)
   Mercury
   Mercuric oxide
   Zinc
   Potassium hydroxide
   Zinc oxide
4. Silver button cell for hearing aids
   Zinc metal
   Mercury metal
   Potassium hydroxide
   Silver oxide
   Zinc oxide

DRY CLEANING ESTABLISHMENTS
Halogenated hydrocarbons—Nonflammable
  Tetrachloroethylene
  Tetracap
  Perclene
  Perchloroethylene
  Carbon tetrachloride
  Tetrachloromethane
  Perchloromethane
Flammables
  Chlorobenzene
  Monochlorobenzene
  Phenyl chloride
  Benzene chloride

ELECTRIC AFTERSHAVE LOTION
Witch hazel
Denatured alcohol
Powdered alum
Boric acid
Glycerine
Water
Perfume

ENAMEL
Quartz
Feldspar
Clay
Soda

Cellulose lacquers

ETHER
Ethyl alcohol (also known as ethyl ether or diethyl ether)

FABRIC SOFTENER
Lauryl pyridinium chloride
Denatured alcohol
Water

FILM CEMENT
Cotton base (nitrocellulose, ethanol, sec.-butyl acetate, methyl ethyl ketone, isopropanol)
Acetone
Methanol
Methyl cellosolve acetate
Ethylene dichloride
May also contain esters, ketones, acetates, dioxane, methylene chloride, butyrolactone

FINGER PAINTS
Bentonite or titanium oxide
Nontoxic colors (vegetable, inorganic, or synthetic)
Preservatives (formaldehyde, benzoic acid salts, dowicil 100, dimethoxane, hydroxyethyl carbinolamine)

FIRE EXTINGUISHERS, FOAM
1. Mechanical foams
   Hydrolyzed protein
   Ethylene glycol
   Cellosolve
   Heavy metal salts (iron, zinc)
   Bactericide
2. Powder foams
   A. Single generator powders with all ingredients in one system
   B. Dual generator powders with aluminum sulfate and soda stabilizer in separate systems
      Sodium bicarbonate
      Aluminum sulfate
      Metal stereates (zinc, aluminum, magnesium)
      Stabilizers (licorice extracts, lignums, glues, egg albumen, dry hydrolyzed protein)
      Clay
3. Liquid foams
   Charge "A"
      Aluminum sulfate-sat. solution
   Charge "B"
      Sodium bicarbonate
      Water
      Stabilizer (licorice extracts, lignums, egg albumen, glues, dry hydrolyzed protein)

FIRE EXTINGUISHERS, LIQUID
1. Chlorobromoethane
2. Halons
   Bromotrifluoromethane
   Dichlorodifluoromethane
   Dichlorotrifluoroethane
   Trichloromonofluoromethane

FIRE EXTINGUISHERS, POWDER
1. Sodium and/or potassium bicarbonate
   Metal stearates (zinc, aluminum, magnesium)
   Flowing agents (mica, alumina, aerogels, tricalcium phosphate)
   Vegetable coloring matter
2. Ammonium monophosphate
   Potassium sulfate
   Flowing agent (aerogel)
   Silicones
   May also contain diammonium phosphate
3. Miscellaneous dry
   Borax compounds
   Clays (bentonite)
   Pitch
   Graphite

FIRE EXTINGUISHERS, SODA ACID
Sodium bicarbonate
Sulfuric acid

FIRE EXTINGUISHERS, WATER BASE
1. Calcium chloride
   Sodium dichromate or chromate
   Water
2. Potassium carbonate
   Ethylene glycol
   Sodium dichromate or chromate
   Water

FIREPLACE COLORS
1. Copper chloride

Coke dust, sawdust
Organic ash

2. Flame colorations
   - Violet:
     - Potassium compounds
     - Rubidium compounds
     - Cesium compounds
   - Blue:
     - Copper chloride, bromide, sulfate
     - Lead compounds
     - Arsenic compounds
     - Selenium compounds
   - Green:
     - Copper sulfate or chloride
     - Copper compounds
     - Thallium compounds
     - Tellurium compounds
     - Barium compounds
     - Molybdenum compounds
     - Antimony compounds
     - Zinc compounds
     - Borates
     - Phosphates
     - Ammonium compounds
   - Red:
     - Strontium nitrate
     - Calcium compounds
     - Lithium compounds
   - Yellow:
     - Sodium compounds

FIREPROOFING COMPOUNDS
Ammonium chloride
Boric acid
Ammonium phosphate

FIREPROOFING FOR TEXTILES
Ammonium phosphate
Ammonium chloride
Water

FIREPROOFING FOR WOOD
Zinc chloride
Ferric chloride
Boric acid
Ammonium phosphate
Water

FLOOR (WOOD) BLEACH
Sodium metasilicate
Sodium perborate
Water

FLOOR POLISH (LIQUID)
Paraffin
Mineral oil

FLOOR POLISH AND WAX (LIQUID)
Yellow beeswax
Ceresin wax
Turpentine
Pine oil
Styrene acrylic latex
Polyethylene waxes
Ammonia
Water

FLY SPRAY
Deodorized kerosene
Pyrethrin
Methyl salicylate

FORMICA
Phenol-formaldehyde resins, which are synthetic resins made by reaction of phenol and formaldehyde

FREON (*See* Refrigerants)

FROST REMOVER
1. Spray Type:
   - Isopropyl alcohol
   - Ethylene glycol
   - Water
   - Propellant (CO_2)
2. Alcohol:
   - Isopropyl alcohol
   - N-propyl alcohol
   - Propylene
   - Ethylene glycol
   - Water
   - May also contain tetrahydrofurfuryl alcohol

FUMIGANTS
Naphthalene
Methyl bromide
Bromomethane
Dihydroacetic acid

FURNACE CEMENT
Sodium silicate
Asbestos fiber
Slate flour, clay, or other filler
Carbon black
Water
May also contain a caustic

FURNACE AND FIREPLACE SOOT DESTROYERS

1. Charcoal
   Potassium nitrate
2. Sodium chloride
   Ammonium chloride
   Copper sulfate
   Wood flour
   Red Iron oxide
   Magnesium carbonate
3. Liquid
   Kerosene
   Copper 6% liquid linoresinate
   Zinc 8% liquid linoresinate
   Ortho dichlorobenzene
   Morpholine
   Petronate
   Butyl alcohol
   Flake naphthalene
   Camphor
   May also contain copper chloride

FURNITURE POLISH (OIL OR WAX)

Mineral oil
Carnauba wax
Stearic acid
Triethanolamine
Perfume
Water

GASKET CEMENTS

1. Epoxy resin
   Polyamide resin
   (These are both liquids and are mixed before using)
2. Batu gum (Scraped)
   Toluene
   Mineral spirits
3. Manila resin
   Denatured alcohol
   May also contain
      Blown vegetable oil
      Isopropyl alcohol
4. Oil-resistant synthetic rubber
   Oil-resistant phenol aldehyde resin
   Acetone

GEAR OILS

Mineral oils
Lubricity agents (chlorinated paraffins, fatty oils)
Oxidation and corrosion inhibitors (barium and calcium phenolates)
Pour depressants and viscosity improvers:
   Polymethacrylate esters
   Polyisobutylenes
Antifoam agents (polysiloxanes)

GERMICIDE (*See* Disinfectant)

GLASS

Silica
Alkali
Lime
Alumina
Lead
Barium
Manganese
Selenium

GLASS CLEANER

Denatured alcohol
Lactic acid
Water

GLUE, GENERAL PURPOSE

1. Dextrin mucilage
   Dextrin
   Borax
   Phenol
   Water
2. Nitrocellulose cements
   Nitrocellulose
   Acetone
   Methyl acetate
   Camphor
3. Polyvinyl acetate wood glue
   Polyvinyl acetate emulsion
   Dibutyl phthalate
4. Gum arabic mucilage
   Gum arabic
   Formaldehyde
   Phenol
   Glycerin
   Oil of wintergreen
   Water
5. Sodium silicate (water glass)
6. Flexible glues:
   Animal glue
   Sucrose
   Polyethylene glycol
   Sodium o-phenylphenate
   Phenol
   Methyl salicylate
   Water
   May also contain alkali

GUN BLUING COMPOUND
Ferric chloride
Antimony chloride
Tannic acid
Boiled linseed oil
Water

GYPSUM
Hydrated calcium sulphate

HAIR CONDITIONER
Deionized water
Mineral oil
Stearalkonium chloride
Hydroxyethyl cellulose
Perfume
Formaldehyde

HAIR SHAMPOO
Water
Tea-lauryl sulfate
Perfume
Triethanolamine
Methylcellulose
Sodium sulfite

HAND CLEANERS, POWDERED
1. Pumice
   Soap
   Sodium phosphate
   Colloidal clay
   Sodium carbonate
2. Borax
   Soap

HAND CLEANERS, WATERLESS
Soap (sodium, potassium, ammonium or triethanolamine)
Solvent (high flash naphtha)
Anionic surfactant
May also contain mild abrasive, water, cream base, and perfume

HAND LOTION
Boric acid
Glycerine
Lanolin, anhydrous
Petrolatum
Perfume

ICE SUBSTITUTE
Roccal
Trisodium phosphate
Water

INK SPOT REMOVER
Sodium perborate
Water

INSECT REPELLENT
Butopyronoxyl
Diethyltoluamide
Dimethyl phthalate
Ethohexadiol
m-Homomenthyl salicylate
Base available as aerosol, cream, foam, liquid, lotion, stick

JEWELRY CLEANERS
Animal fatty acid
Aqua ammonia
Isopropanol
Dye
Perfume
Water
May also contain cyanide, thiourea, sulfuric acid, and polyethylene glycols

LACQUER
Nitrocellulose
Cellulose acetate
Solvents (anhydrous alcohol, ethyl acetate, butyl acetate, benzol, toluol)
Softeners (amyl, ethyl, butyl phthalates)

LAUNDRY BLUING
Ultramarine blue
Sodium bicarbonate
Corn syrup

LAUNDRY DETERGENT (ENZYME ACTIVE)
Pancreatin
Sodium chloride
Sodium carbonate
Sodium alginate

LAWN AND TURF FUNGICIDE
Thiram
Mercurous chloride
Mercuric chloride

LEATHER PRESERVATIVE
Neatsfoot oil (obtained by boiling the feet of cattle or sheep in water, skimming the oil from the surface, and pressing through a filter.

Contains about 80% oleic acid)
Castor oil

LINOLEUM
Fabric impregnated and covered with a compound of blown linseed oil, ground cork or wood flour, and resin or other gum. A filler of lithopone may also be used.

LIGHTER FLUID
Petroleum hydrocarbons (naphthas)
Perfume

LUCITE® (*See* Acrylic Resins)

LUMINOUS PAINT
Paints that make the surface visible in the dark. Two basic types are manufactured. Permanently luminous contains a radioactive substance and obtains energy from decomposition of radioactive pigment.
The second type, phosphorescent paints, obtain their energy from outside light sources that are absorbed, and then emit themselves in the form of photons of light.
Phosphorescent pigments are usually combinations of zinc and cadmium sulphides (for yellow and orange), while violet and blue pigments are combinations of calcium and strontium sulphides.

LYSOL® (*See* Disinfectants)

MARINE GLUES
1. Dry casein glue
   Casein
   Calcium hydroxide
   Sodium fluoride
   Kerosene
2. Epoxy adhesive
   Epoxy resin
   Polyamide resin

NOTE: These two are mixed together prior to use.
3. Resorcinol-formaldehyde resin

MASTIC
The gum exudation of the tree *Pictacia lentiscus,* native to Mediterranean countries. It is used and found in lacquers, varnishes, and adhesives.

MATCHES
Animal glue
Potassium chlorate
Sulfur
Wood resin
Silica flour (or ground glass)
Phosphorus sequisulfide (strike anywhere matches only

MATCH-STRIKING SURFACE
Binder of synthetic resins, latex, or a natural gum arabic
Red phosphorus
Silica flour or ground glass

MERCERIZED COTTON
Cotton yards treated with sodium hydroxide

MERCURY CELL BATTERIES
(*See* Dry Cell Batteries)

METAL CEMENTS
Inert ingredients (largest percentage)
Vinyl plastic
Ketone solvents
Diethyl phthalate
Butyl acid phosphate

MICARTA (*See* Formica)

MICROFILM CEMENT
Cotton base:
   Nitrocellulose
   Ethanol
   Isopropyl alcohol
   Toluene
Acetic acid

MIMEOGRAPH CORRECTION FLUID
1. Ether type
   Diethyl ether
   Plasticizers
   Ethanol
   Nitrocellulose
   Pigments, dyes, or toners
2. Nonether type
   Ethyl acetate
   Ethanol
   Plasticizer
   Nitrocellulose
   Pigments, dyes, or toners

MODEL AIRPLANE FUEL
1. Alcohol type

Methanol
Nitromethane
Ucon synthetic oil or castor oil

2. Diesel type
   Kerosene
   Castor oil
   Ether

MODEL CEMENT
Nitrocellulose
Acetone
Cellulose acetate
Isopropanol
Dibutyl phthalate
Hexane
Ethyl
Toluol
Naphtha
Ethanol
Camphor
Polystyrene

MODELING CLAY
Kaolin
Gypsum
Petrolatum
Wax
Castor oil
Vegetable dyes

MOSQUITO REPELLENT
Eucalyptus oil
Talc
Corn starch

MOTH PROOFING COMPOUND
Ammonium selenate
Water

MOUTH WASH ASTRINGENT (LIQUID)
Sodium bicarbonate
Borax
Zinc chloride
Menthol crystals
Denatured alcohol
Glycerine
Water
Food coloring

OPALON (*See* Formica)

OILCLOTH
Cotton, jute, or hemp covered with turpentine and resin

OVEN CLEANERS
1. Potassium or sodium hydroxide
   Starch, glue, etc.
   Nonionic synthetic surfactant
2. Sodium hydroxide
3. Sodium orthosilicate
   Whiting (chalk)
   Cellulose
   N-Acetyl ethanolamine
   Butyl cellosolve
   Sodium xylene sulfonate
   Water
4. Spray-type:
   Amines
   Detergents
   Ether-type solvents
   Propellants
   Sodium hydroxide
   Thickeners

OVEN CLEANER (POWDER)
Trisodium phosphate
Soda ash
Sodium perborate
Powdered soap

PAINTS
Paints are mixtures of binders, also called filmogens, and pigments. The binder provides continuity and the pigment provides the color. Varnish is usually a blend of drying oil and resin (*See* Varnish, below), while conventional oil-based paints use binders of vegetable oil. Water thinned paints usually utilize solubilized linseed oil or casein as the binder. Enamel dries relatively harder and smoother than glossier paints and lacquer has come to mean a paint based on cellulose nitrate or cellulose acetate butyrate.

If driers are added during the cooking process, the end product is referred to as boiled oil. If oxygen is bubbled through the product during cooking, the final result is referred to as blown oil.

Recent inovations in processing have resulted in the marketing of heat resistant and fire retardant paints. In this regard it should be recognized that most paint films probably contain less of a combustible product than does wood. The premise is that the paint will swell and bubble (the intumescent effect) and this bubbling acts as an insulator to the fire. There are, however, limitations on the time of flame or hostile fire retardancy with these paints. Silicone paints probably are the most heat-resistant.

What follows is an analysis of some different types of paints which are most likely to be encountered:

1. Aluminum, bronze and other metallic paints
   - Aluminum pigment
   - Petroleum resin
   - Vegetable oil varnish
   - Linseed oil
   - Aromatic thinners
   - Mineral spirits
   - (Bold bronze contains copper or copper-tin alloy pigment)
2. Varnish
   - Alkyd resin
   - Vegetable oil varnish
   - Mineral spirits
   - Fillers and pigments
   - Polyurethane resin
   - Phenolic resin
3. Asphalt paint, including screen enamel and stove enamel
   - Asphalt or gilsonite
   - Oil or varnish
   - Mineral spirits
   - Carbon black
4. Stains, penetrating, or wiping
   - Gilsonite or asphalt
   - Aniline dyes
   - Vegetable oil varnish
   - Alkyd resin
   - Mineral spirits
5. Flat wall paint oil, primers, and sealers
   - Titanium dioxide
   - Titanium calcium
   - Lithopone
   - Alkyd resin or oil varnish
   - Mineral spirits
6. Semi-gloss, satin or egg-shell enamel
   - Titanium dioxide
   - Titanium calcium
   - Lithopone
   - Mineral spirits
   - Alkyd resin or oil varnish
   - Inert filler pigments
7. Gloss enamel and floor enamels
   - Titanium dioxide
   - Titanium calcium
   - Inert pigments
   - Mineral spirits
8. Latex wall paint, texture coatings, and fillers
   - Titanium dioxide
   - Inert filler pigments
   - Polyvinyl acetate elastomer
   - Acrylic lastomer
   - Styrene butadiene elastomer
   - Lithopone
   - Emulsifying agents
   - Water
   - Vegetable oils or resins
   - Ethylene glycol
9. Latex house paint
   - Titanium dioxide
   - Inert filler pigments
   - Polyvinyl acetate elastomer
   - Acrylic elastomer
   - Styrene butadiene elastomer
   - Ethylene glycol
   - Zinc oxide
   - Alkyd resin
   - Chlorinated phenols
   - Water
   - Vegetable oils or resins
   - May also contain mercury fungicides
10. House paint, trim paint, exterior
   - Titanium dioxide
   - Titanium calcium
   - Zinc oxide
   - White lead (not usually found in latex paints although exterior latex paints may contain mercury fungicides)
   - Inert filler pigments
   - Linseed oil
   - Alkyd resins
   - Mineral spirits

PAINT, ANTIALGAE

Cuprous oxide

Arsenic oxide
Copper sulfate pentahydrate
Mercury soap

PAINT, ANTICORROSION

Rosin
Pine oil
Coal tar
Petroleum ether
May also contain ammonium hydroxide, arsenic, cuprous oxide, ethyl alcohol, kerosene, mercuric oxides, metallic soap, methylene chloride, paraffin, plastic thickener, and rust retardant

PAINT BRUSH CLEANERS

1. Benzene (or toluene or xylene)
   Acetone
   Methyl alcohol
   Naphtha or stoddard solvent
2. Kerosene
   Oleic acid
   Ammonia

PAINT DRIERS

Vanadium compounds
Manganese compounds
Zinc compounds
Iron compounds
Cobalt compounds
Potassium hydroxide
Lead naphthenate
Lead octoate
Lead tallate

PAINT AND VARNISH CLEANERS

1. Liquid
   A. Soap
      Anionic surfactant
      Water
   B. Sulfated fatty alcohol
      Ammonia
      Water
   C. Soap
      Kerosene
      Colloidal clay
      Water
2. Paste
   Soap
   Kerosene
   Trisodium phosphate
   Water
   Perfume
3. Powdered
   Trisodium phosphate
   Soda ash
   Soap
   May also contain sodium bicarbonate, ammonium chloride

PAINT AND VARNISH REMOVERS

1. Flammable
   Benzene
   Methanol
   Acetone
   Paraffin wax
   Toluene
   Ethyl cellulose
2. Non-flammable
   Methylene chloride
   Methanol
   Toluene
   Methyl or ethyl cellulose
   Paraffin wax
3. Stripping agents/paint removers
   Phenols
   Cresols
   Methylene chloride
   Organic acid catalyst (acetate, formate)
   Concentrated alkali

PAINT THINNERS

Acetone
Aliphatic petroleum solvents
Aromatic hydrocarbon solvents (xylene, toluene)
Cellulosolve acetate
Diacetone alcohol
Ethyl alcohol
Ethyl amyl ketone
Isopropyl alcohol
Methyl ethyl ketone
Methyl isobutyl ketone
Naphthol
Naphthenes
Toluene
Exlene

PAPER FIREPROOFING COMPOUND

Ammonium sulfate
Boric acid
Borax
Water

PARAFFIN
Methane
Naphtha
Benzene
NOTE: Paraffin is flame resistant and is used in treating paper and textiles

PASTE, LIBRARY
1. Corn, wheat or potato dextrin
Glucose
Phenol or sodium O-phenylphenate
Oil of wintergreen
Glycerin
Water
2. Starch gum
Glycerin
Defoamer
Formaldehyde
Phenol
Water

PATCHING CEMENTS
1. Portland cement (fused lime and clay)
Calcium hydroxide
Sodium chloride
Titanium dioxide
2. Wall patching plaster
Plaster of paris
Dextrin
Pumice powder
3. Plastic putty
Gum solution (16 lb. of rosin/gal methyl acetone)
Glycerin
Butyl alcohol
Whiting
Wood flour
Dope solution (1 lb. movie film (cellulose acetate or nitrate) per gallon of methyl acetone)

PENETRATING OILS
Mineral oils
Kerosene
May also contain surface active ingredients, fatty oils and fatty acids, xylene, tributyl phosphate, butyl alcohol, chlorinated hydrocarbons

PHENOLITE (*See* Formica)

PIPE JOINT CEMENT
1. Chalk
Kaolin
Linseed oil
Litharge
2. Black strap molasses
Amorphous graphite
Vermiculite
Bentonite
Lithopone
Sodium pentachlorophenate

PLASTERBOARD (*See* Gypsum)

PLASTIC CEMENT
1. Polyvinyl acetate
Clay (kaolin)
Polyvinyl alcohol
Sodium O-phenylphenate
2. Nitrocellulose cement
Nitrocellulose
Acetone
Methyl acetate
Camphor
3. Rubber cement
Hexane
Rubber
4. Polystyrene
Benzene
Toluene
Acetone
Naphtha

PLASTIC WOOD
Wood cellulose dissolved in ether. It may also have phenoli resin to accelerate curing.

PLAY DOH®
Nontoxic salts
Cereal products
Highly purified hydrocarbon distillates

PLEXIGLAS (*See* Acrylic Resins)

POLYURETHANE (*See* Urea Formaldehyde Resins)

RAYON
Artificial silk textile fibers or yarns made from cellulose nitrate, cellulose acetate or cellulose derivatives. Viscose rayon is made by treating cellulose with caustic soda and then with carbon disulphide to form cellulose xanthate

READY MIXED JOINT COMPOUND
- Plasticized polyvinyl acetate emulsion
- Diethylene glycol nonoethyl ether
- Phenyl mercury acetate
- Mica
- Asbestos
- Calcium carbonate
- Water

REFRIGERANTS (Usually Identified by Number)
- Ammonia
- Ethyl ether
- Carbon dioxide
- Sulphur dioxide

Number	*Name*
R-11	Trichloromonofluoromethane
R-12	Dichlorodifluoromethane
R-22	Monochlorodifluoromethane
R-500	Azeotropic mixture of 73.8% of R-12 and 26.2% of R-152a
R-502	Azeotropic mixture of 48.8% of R-22 and 51.2% of R-115
R-503	Azeotropic mixture of 40.1% of R-23 and 59.5% of R-13
R-504	Azeotropic mixture of 48.2% of R-52 and 51.8% of R-115
R-717	Ammonia (NH_3)

Refrigerants are nonexplosive, noncorrosive, nonflammable and usually operate under low pressure (low boiling point)

ROCKOUSTILE (*See* Vermiculite)

ROOFING CEMENT
- Petroleum asphalt
- Petroleum solvents
- Asbestos fiber
- Limestone, clay, or other filler

RUBBER CEMENT
1. Hexane, gasoline, benzol, naphtha, alcohol (solvents)
   Rubber-natural, SBR
2. Butadiene styrene latex
   Rosin gum emulsion
   Water
   Ammonium hydroxide
   Sodium O-phenylphenate
   Sodium pentachlorophenate
   Cunilate 2778-I
   Colloid 581-B
3. Centrifuged natural latex
   Black color (carbon)
   Ethylene glycol
   Dispersed sulfur
   Trisodium phosphate
   Creosote
   Stabilizer
   Ammonia
   Inert ingredients (earths, flours, etc.)

RUBBING ALCOHOL
- Denatured alcohol
- Glycerine
- Castor oil

RUG AND CARPET CLEANER (LIQUID)
- Denatured alcohol
- Vinegar
- Lauryl pyridinium chloride

RUG AND UPHOLSTERY CLEANERS
1. Synthetic detergents
   Fatty alcohol sulfate
   Trisodium phosphate
   Alcohol
   Water
2. Alkaline cleaners
   - Soda soap
   - Trisodium phosphate
   - Sodium carbonate
   - Borax
   - Naphthalene
   - Oils
3. Soap-solvent combinations
   - Butyl cellosolve
   - Ethylene dichloride
   - Oleic acid
   - Ammonia
   - Triethanolamine
   - Soap
   - Borax
   - Denatured alcohol
   - Isopropyl alcohol
   - Water
4. Upholstery
   - Perchlorethylene
   - Petroleum solvent (stoddard)
   - Alkaline salts
   - Butane-freon propellant (for sprays only)
   - Water
5. Absorbant (for rugs)

A. Sawdust
Trichloroethylene
B. Buckwheat or bentonite
Petroleum distillate (light)
Wood flour

SHAVING CREAM
Water
Stearic acid
Triethanolamine
Isobutane
Sodium lauryl sulfate
Propane
Stearyl alcohol
Perfume/fragrance

SHELLAC
The lac secretion produced by *Tachardia lacca,* an insect that lives on various trees of southern asia. Shellac varnish is then made by cutting the resin in alcohol.

SHOE CEMENT
1. Hexane, gasoline, naphtha (solvents)
Rubber-natural, SBR
2. Cellulose nitrate
Acetone
Aromatic hydrocarbon solvents
3. Butadiene styrene latex
Rosin gum emulsion
Water
Casein
Ammonium hydroxide
Sodium O-phenylphenate
Sodium pentachlorophenate
Cunilate 2778-I
Colloid 581-B
4. Polyvinyl butyral
Rosin
Castor oil, hydrogenated
Oleamide, stearamide, palmitamide mixture
Titanium dioxide
Clay (kaolin)
Alcohol

SHOE AND BOOT WATERPROOFING COMPOUND
Silicone oil
Stoddards solvent

SILVER CLEANER
Water
Stearic acid
Soda ash
Trisodium phosphate
Diatomaceous earth

SKI WAX
Tar (wood)
Diglycol sterate
Carnauba wax

SMOKE FLUID FOR TOY TRAINS
Pure white mineral oil
Hexahydrothymol
Fractol (mineral oil)

SNOW AND ICE MELTING COMPOUND
Rock salt
Ammonium chloride
Magnesium sulfate

SPACKLING COMPOUND
Calcium sulfate
Mica
Glue

STAINLESS STEEL CLEANER
Sulfamic, citric, tartaric acid
Sodium chloride
Anionic synthetic surfactant
Corrosion inhibitor (sodium acid phosphates, borax, etc.)

STEAM
Water vapor

STEEL
Iron with carbon chemically dissolved into it

STERNO
Ethyl alcohol
Denatured alcohol
Methanol

SUEDE (Napped Leather)
A soft leather made from calf, kid, or cowhide splits or from sheepskin. The leather is processed on a staking machine until it becomes soft after which it is polished (or buffed) on an abrasive wheel

SUNTAN LOTIONS AND PREPARATIONS
1. Cream (Emulsion)
A. Water

Fats, oils, waxes
Emulsifiers
Humectants
Preservatives
Fragrances
Sunscreen agents

B. Monoglyceryl P-aminobenzoate
Mineral oil
Sorbitan monosterate
Polyoxyethylene sorbitan monosterate
Water

2. Gels and oils
Oils, usually mineral
Emollient
Thickeners
Sunscreen agents
Preservatives
Coloring
Fragrances
3. Lotion (alcoholic)
Menthyl anthranilate
Propylene glycol ricinoleate
Glycerol
Alcohol
Water
4. Oil
2-Ethyl hexyl salicylate
Sesame oil
Mineral oil
5. Indoor tanning lotions
Lanolin (deodorized)
Vegetable oil/(sesame oil
Mineral oil
Water
Fragrances
Tanning agents (dihydroxy acetone)
Sunscreen agents

TANTALUM

A white, lustrous metal resembling platinum which is acid-resistant and classed as a noble metal. Melting point is about 2850° Centigrade. Tantalum is dissolvable in a mixture of nitric acid and hydrofluoric acids and will also react with chlorine above 175° Centigrade.

Tantalum is used as filament material in electric light bulbs and as heating coils in toasters and other similar appliances. The metal becomes incandescent at 1700° Centigrade which is 400° Centigrade below tungsten so tantalum burns cooler.

TEFLON CLEANER

Sodium perborate
Sodium metasilicate
Inorganic salts
Synthetic organic detergents

TERMITE REPELLENT

Para dichlorobenzene
Denatured alcohol

TEXTOLITE (*See* Formica)

THERMOPLASTIC

1. Polyvinyl acetate emulsion
Diethyl phthalate
Silicone defoamer
Methylcellulose
Water
Phenol
2. Polyvinyl acetate emulsion
Dibutyl phthlate
3. Polyvinyl butyral
Rosin
Castor oil (hydrogenated)
Oleamide, palmitamide, stearamide mixture
Titanium dioxide
Clay (kaolin)
4. Butadiene styrene latex
Rosin gum emulsion
Casein
Ammonium hydroxide
Sodium O-phenylphenate
Sodium pentachlorophenate
Cunilate 2778-I
Colloid 581-B
Water

THINNERS (*See* Paint Thinners)

TILE ADHESIVES

1. Rubber/resin emulsion
Resin tackified rubbers
Rosin gum emulsion
Butadiene styrene latex
Water
Casein
Ammonium hydroxide
Sodium O-phenylphenate
Cunilate 2778-I
Colloid 581-B

2. Rubber/resin solvent
   Inert rubbers and resins
   Inert mineral fillers
   Petroleum solvents or alcohol solvents
3. Asphalt emulsion
   Inert asphalt
   Inert mineral fillers
   Water
4. Water base
   Waste sulphite liquor
   Preservatives
   Inert mineral filler
5. Epoxy resin
   Polyamide resin
   (These are mixed just prior to application)
6. Butadiene styrene latex
   Rosin gum emulsion
   Casein
   Ammonium hydroxide
   Sodium O-phenylphenate
   Sodium pentachlorophenate
   Cunilate 2778-I
   Colloid 581-B
   Water
   2,2-Methylenebis (4-methyl-6-tertiary-butylphenol)

TIRE CLEANERS

1. Potassium hydroxide
   Metasilicates
   Alkyl aryl sulfonate
   Water
2. Aerosol whitewall cleaner
   Anionic detergents
   Phosphates
   Aliphatic solvents
   Aerosol propellants
3. Trisodium phosphate or sodium tripolyphosphate
   Metasilicate
   Alkyl aryl sulfonates
   Butyl cellosolve
   EDTA
   Perfume
   Fluorescein

TUNGSTEN

A heavy white metal with a specific gravity of 19.6 and a melting point of 6100° Fahrenheit obtained in powder form by reduction of the oxide WO_3 in a stream of hydrogen.

TURPENTINE

An oil obtained by steam distillation of the oleoresin which exudes when various conifer trees are cut. Wood turpentine, also known as spirits of turpentine, is obtained from waste wood, chips, or sawdust by steam extraction.

Terpene alcohol, or methyl pinene, is produced by condensing the beta pinene of gum turpentine with formaldehyde.

TYPEWRITER CLEANERS

1,1,1-Tric loroethane (methyl chloroform)

UNDERCOATINGS

1. Pigment
   Iron oxide
   Silicates
   Lead or lead salts
   Zinc and zinc salts
   Titanium dioxide
2. Vehicle
   Alkyd resin
   Phenolic resin
   Fish oil
   Vegetable oil
   Aromatic hydrocarbons
   Aliphatic hydrocarbons

UREA-FORMALDEHYDE RESINS

Urea is also known as Carbamide. Urea resins are made by condensation of urea or thiourea with formaldehyde. As a gloss finish it may include polyhydric alcohol, stearic acid and mineral spirits. Polyurethane foams are a class of thermosetting plastics set in molds. The basic components include isocyanates, polyethers, castor oils, amines, carboxylic acid and water. After involvement in a hostile fire the investigator may encounter a sweet or fishy odor in the residue.

VARNISH (*See* Paints)

VEGETABLE GARDEN DUSTS

Carbaryl
Malathion
Captan

VERMICULITE
A foliated mineral used in plasters and insulation and as a filler in caulking compounds. It is usually yellowish to brown. Plaster made with 60% vermiculite, 30% plaster of Paris and 10% asbestos will withstand red heat without disintegrating.

WALLBOARD (*See* Gypsum)

WALL PAPER REMOVER
Tergitol
Ethylene glucol monobutyl ether
Water

WATER REPELLENT
Mineral salts
Wax emulsions
Aluminum acetate
Zirconium acetate

WATER SOFTENER
Sodium silicate
Soda ash
Sodium hydroxide
Barium carbonate
Alum
Zeolite

WEED AND BRUSH KILLER
Ammonium sulfamate

WINDSHIELD WASHER ANTIFREEZE
Methyl alcohol

WINDSHIELD WASHER SOLVENT
Isopropyl alcohol

WOOD PRESERVATIVES
Gum resins
Denatured alcohol
Naphthenic acid
Copper hydrate
Mineral spirits
Pentachlorophenol

ZIPPER LUBRICANT
Petroleum wax in perchlorethylene base
Lubricant may also contain carnauba wax, lacquer solvents
Microcrystalline waxes
Petroleum oils (naphthenic)
Pour point depressant (polymeric type)
Trichloroethylene
Varnish

ZONOLITE (*See* Vermiculite)

7

REFERENCE TABLES

Table 7-1: Melting Points of Nonmetallic Elements (° F)

Helium	−456*	Argon	−309	Phosphorus	111
Hydrogen	−434	Krypton	−272	Iodine	236
Neon	−416	Xenon	−220	Sulphur	235
Fluorine	−367	Chlorine	−151	Silicon	2588
Oxygen	−362	Bromine	+19	Carbon	>6500
Nitrogen	−346				

*At 25 atm.

Baumeister T: Marks' Standard Handbook for Mechanical Engineers, 8th ed. New York, McGraw-Hill, pp. 4-5, 1978, Reprinted by Permission.

Table 7-2: Melting Points of Various Solids (° F)

Alloys:		Fusible alloys:		Enamel colors	1760
Bismuth solder	200-262	33 Bi + 33 Pb + 33 Sn	250	India rubber	257
Brass and bronze (about)	1650	18 Bi + 36 Pb + 46 Sn	305	Paraffin	129
80 Cu + 20 Zn	1845	10 Bi + 40 Pb + 50 Sn	324	Phosphorus (white or yellow)	111
50 Cu + 50 Zn	1615	Tin solder	275-350	Porcelain	2820
20 Cu + 80 Zn	1300	Blast-furnace slag	2370-2600	Potassium	147
Delta metal	1742	Borax	1040	Sodium	208
20 Sn + 80 Pb	530	Cast iron, gray	2460-2550	Spermaceti	120
50 Sn + 50 Pb	400	Cast iron, white	1920-2010	Stearine	122
80 Sn + 20 Pb	388	Chlorides:		Steel	2370-2550
		Calcium	1422	Wrought iron	2460-2640
		Potassium	1454		
		Sodium	1479		
		Zinc	541		

Baumeister T: Marks' Standard Handbook for Mechanical Engineers, 8th ed. New York, McGraw-Hill, pp. 4-5, 1978, Reprinted by Permission.

Table 7-3: Freezing Points of Liquids at Atmospheric Pressure (°F)

Ammonia	−107.8	Linseed oil	−4
Aniline	20.8	Mercury	−38.8
Bensol	41.9	Methyl alcohol	−144.2
Carbon bisulphide	−168.1	Rapeseed oil	25.7
Carbon dioxide	−110.2	Turpentine	<−75
Chloroform	−82.3	Sulphuric acid	−105
Calcium chloride (sat sol)	−40	Salt (NaCl) sol, sat	−0.4
Ether	−180	Sea water	27.5
Ethyl alcohol	−174.6	Toluene	−149
Glycerin	64		
Naphthalene	176		

Baumeister T: Marks' Standard Handbook for Mechanical Engineers, 8th ed. New York, McGraw-Hill, pp. 4-5, 1978, Reprinted by Permission.

Table 7-4: Boiling Points at Atmospheric Pressure (°F)

Zinc	1665	Naphthalene	424
Sulphur	832	Aniline	364
Mercury	675	Calcium chloride (sat sol)	356
Linseed oil	549	Turpentine	320
Paraffin	572	Toluene	231
Glycerin	554	Sodium chloride (sat sol)	226.4
Phosphorus	536	Helium	−452

Baumeister T: Marks' Standard Handbook for Mechanical Engineers, 8th ed. New York, McGraw-Hill, pp. 4-5, 1978, Reprinted by Permission.

Table 7-5: Boiling Points of Hydrocarbons (°F)

Name	Chemical formula	°F
	Paraffins or Alkanes: C_nH_{2n+2}	
Methane	CH_4	−258.9
Ethane	C_2H_4	−127.5
Propane	C_3H_8	−44.2
Butane	C_4H_{10}	31.3
Pentane	C_5H_{12}	96.9
Hexane	C_6H_{14}	155.6
Heptane	C_7H_{16}	209.1
Octane	C_8H_{18}	258.2
Nonane	C_9H_{20}	303.2
Decane	$C_{10}H_{22}$	345.2
	Olefine or Alkenes: C_nH_{2n}	
Ethylene	C_2H_4	−155.0
Propylene	C_3H_6	−53.7
1-Butene	C_4H_8	20.8
1-Pentene	C_5H_{10}	86.0
	Diolefine or Dienes: C_nH_{2n-2}	
1,3 Butadiene	C_4H_6	24.1
Isoprene	C_5H_8	93.3
	Acetylenes or Alkines: C_nH_n	
Acetylene	C_2H_2	−119.2
	Aromatic Series	
Benzene	C_6H_6	176.1
Toluene	C_7H_8	231.2
Ethylbenzene	C_8H_{10}	277.1
Naphthalene	$C_{10}H_2$	424.2

Baumeister T: Marks' Standard Handbook for Mechanical Engineers, 8th ed. New York, McGraw-Hill, pp. 4-5, 1978, Reprinted by Permission.

Table 7-6: Composition of the Atmosphere (% by Volume)

Gas	%
Nitrogen	78.03
Oxygen	20.99
Argon	0.94
Carbon dioxide	0.03
Hydrogen	0.01
Neon	0.00123
Helium	0.0004

Brady: Materials Handbook: an Encyclopedia for Purchase Agents, Engineers and Foremen, 7th ed, New York, McGraw-Hill, p 799, 1951. Reprinted by Permission.

Table 7-7: Coefficients of Expansion

Mean values of 10,000s' between 32 and 212°F; 0 and 100°C

Metals		Glass: (cont'd)	
Aluminum bronze	0.094	Plate and crown	0.050
Brass, cast	0.104	Flint	0.044
Brass, wire	0.107	Pyrex	0.018
Bronze	0.100	Granite	0.04-0.05
Constantan (60 Cu, 40 Ni)	0.095	Graphite	0.044
German silver	0.102	Gutta percha	0.875
Iron:		Ice	0.283
Cast	0.059	Limestone	0.023-0.05
Soft forged	0.063	Marble	0.02-0.09
Wire	0.080	Masonry	0.025-0.050
Magnalium (85 Al, 15 Mg)	0.133	Paraffin:	
Phosphor bronze	0.094	32 F-61 F	0.592
Solder	0.134	61 F-100 F	0.724
Speculum metal	0.107	100 F-120 F	2.612
Steel:		Porcelain	0.02
Bessemer, rolled hard	0.056	Quartz:	
Bessemer, rolled soft	0.063	Parallel to axis	0.044
Nickel (10% Ni)	0.073	Perpend. to axis	0.074
Type metal	0.108	Quartz, fused	0.0028
Other Materials		Rubber	0.428
Bakelite, bleached	0.122	Vulcanite	0.400
Brick	0.053	Wood (‖ to fiber):	
Caoutchouc	0.372	Ash	0.053
Carbon—coke	0.030	Chestnut and maple	0.036
Cement, neat	0.060	Oak	0.027
Concrete	0.080	Pine	0.030
Ebonite	0.468	Across the fiber:	
Glass:		Chestnut and pine	0.019
Thermometer	0.045	Maple	0.027
Hard	0.033	Oak	0.030

Baumeister T: Marks' Standard Handbook For Mechanical Engineers, 8th ed. New York, McGraw-Hill, p 4-64, 1978. Reprinted by Permission.

Table 7-8: Thermal Conductivities of Miscellaneous Solid Substances (Values of *k* are to be regarded as rough average values for the temperature range indicated)

Material	Bulk density, lb per cu ft	Temp, ° F	*k*
Asbestos board, compressed asbestos and cement	123.	86	0.225
Asbestos millboard	60.5	86	0.070
Asbestos wool	25.	212	0.058
Ashes, soft wood	12.5	68	0.018
Ashes, volcanic	51.	300	0.123
Carbon black	12.	133	0.012
Cardboard, corrugated	—	—	0.037
Celluloid	87.3	86	0.12
Cellulose sponge, du Pont	3.4	82	0.033
Concrete, sand and gravel	142.	75	1.05
Concrete, cinder	97.	75	0.41
Charcoal, powder	11.5	63	0.029
Cork, granulated	5.4	23	0.028
Cotton wool	5.0	100	0.035
Diamond	151.	70	320.
Earth plus 42% water	108.	0	0.62
Fiber, red	80.5	68	0.27
"Flotofoam" (U.S. Rubber Co.)	1.6	92	0.017
Glass, pyrex	139.	200	0.59
Glass, soda lime	—	200	0.59
Graphite, solid	93.5	122	87.
Gravel	116.	68	0.22
Gypsum board	51.	99	0.062
Ice	57.5	—	1.26
Kaolin wool	10.6	800	0.059
Leather, sole	62.4	—	0.092
Mica	122.	—	0.25
Pearlite, Arizona, spherical shell of siliceous material	9.1	112	0.035
Polystyrene, expanded "Styrofoam"	1.7	—	0.021
Pumice, powdered	49.	300	0.11
Quartz, crystal, perpendicular to *C*-axis	—	−300	12.5
		0	4.3
		300	2.3
Quartz, crystal, parallel to *C*-axis	—	−300	25.0
		0	8.3
		300	4.2
Rubber, hard	74.3	100	0.092
Rubber, soft, vulcanized	68.6	86	0.08
Sand, dry	94.8	68	0.188
Sawdust, dry	13.4	68	0.042
Silica, fused	—	200	0.83
Silica gel, powder	32.5	131	0.049
Soil, dry	—	68	0.075
Soil, dry, including stones	127.	68	0.30
Snow	7-31	32	0.34-1.3
Titanium oxide, finely ground	52.	1000	0.041
Wool, pure	5.6	86	0.021
Zirconia grain	113.	600	0.11

Table 7-8: Thermal Conductivities of Miscellaneous Solid Substances (cont'd.)

Material	Bulk density, lb per cu ft	Temp, ° F	*k*
Woods, oven dry, across grain†:			
Aspen	26.	85	0.069
Bald cypress	24.	85	0.063
Balsa	10.	85	0.034
Basswood	24.	85	0.058
Douglas Fir	29.	85	0.063
Elm, rock	48.	85	0.097
Fir, white	26.	85	0.069
Hemlock	29.	85	0.066
Larch, western	36.	85	0.078
Maple, sugar	43.	85	0.094
Oak, red	42.	85	0.099
Pine, southern yellow	35.	85	0.078
Pine, white	25.	85	0.060
Red cedar, western	21.	85	0.053
Redwood	25.	85	0.062
Spruce	21.	85	0.052

*The thermal conductivity of different materials varies greatly. For metals and alloys *k* is high, while for certain insulating materials, such as glass, wood, cork, and kapok, it is very low. In general *k* varies with the temperature, but in the case of metals, the variation is relatively small. With most other substances, *k* increases with rising temperatures, but in the case of many crystalline materials, the reverse is true.

† With heat flow parallel to the grain, *k* may be 2 to 3 times that with heat flow perpendicular to the grain; the values for wood are taken chiefly from MacLean JD: *Trans. ASHRAE,* 47, 1941, p 323.

Baumeister T: Marks' Standard Handbook For Mechanical Engineers, 8th ed, New York, McGraw-Hill, p 4-64, 1978. Reprinted by Permission.

Table 7-9: Thermal Conductivities for Building Insulations

Material	Bulk density, lb per cu ft	Temp, °F	k
Balsam wool, blanket	3.6	70	0.021
Cabot's Quilt, eel-grass	15.6	86	0.027
Glass wool, blanket	3.25	100	0.022
Hairfelt, blanket	11.0	86	0.022
Insulating boards, Insulite, Celotex, etc.	12-19	100	0.027-0.031
Kapok, "DryZero," blanket	1.6	75	0.019
Redwood bark, loose, shredded, "Palco Bark"	4.0	100	0.025
Rock wool, loose	7.	117	0.024
Sil-O-Cel powder	10.6	86	0.026
Vermiculite, loose, "Zonolite"	8.2	60	0.038

Baumeister T: Marks' Standard Handbook For Mechanical Engineers, 8th ed. New York, McGraw-Hill, p 4-64, 1978. Reprinted by Permission.

Table 7-10: Thermal Conductivities of Materials for Refrigeration and Extreme Low Temperatures

Material	Bulk density, lb per cu ft	Temp, ° F	*k*
Corkboard	6.9	100	0.022
		−100	0.018
		−300	0.010
Fiberglas with asphalt coating (board)	11.0	100	0.023
		−100	0.014
		−300	0.007
Glass blocks, expanded, "Foamglas"	10.6	100	0.036
		−100	0.033
		−300	0.018
Mineral wool board, "Rockcork"	14.3	100	0.024
		−100	0.017
		−300	0.008
Rubber board, expanded, "Rubatex"	4.9	100	0.018
		−100	0.015
		−300	0.004
Silica aerogel, powder, "Santocel"	5.3	100	0.013
		0	0.012
		−100	0.010
Vegetable fiberboard, asphalt coating	14.4	100	0.028
		−100	0.021
		−300	0.013
Foams:	2.9	−100	0.015
Polystyrene[a]			
Polyurethane[b]	5.0	−100	0.019

[a]Test space pressure, 1.0 atm; $k = 0.0047$ at 10^{-5} mmHg.
[b]Test space pressure, 1.0 atm; $k = 0.007$ at 10^{-3} mmHg.

Baumeister T: Marks' Standard Handbook For Mechanical Engineers, 8th ed. New York, McGraw-Hill, p 4-64, 1978. Reprinted by Permission.

Table 7-11: Strength and Related Properties of Wood at 12% Moisture Content (Average Values from Tests on Clear Pieces 2″ × 2″ in Section Per ASTM D-143)

Kind of wood	Specific gravity, ovendry volume	Density at 12% m.c. lb/ft³	Shrinkage, % from green to ovendry condition based on dimension when green: Rad.	Shrinkage: Tan.	Static bending: Modulus of rupture, lb/in²	Static bending: Modulus of elasticity, ksi	Max crushing strength parallel to grain, lb/in²	Compression perpendicular to grain at proportional limit, lb/in²	Tensile strength perpendicular to grain, lb/in² †	Impact bending, height of drop in inches for failure with 50-lb hammer	Shear strength parallel to grain, lb/in²	Hardness perpendicular to grain, avg of R and T
Hardwoods												
Ash, white	0.64	42	4.9	7.9	15,400	1,770	7,410	1,160	940	43	1,950	1,320
Basswood	0.40	26	6.6	9.3	8,700	1,460	4,730	370	350	16	990	410
Beech	0.67	45	5.1	11.0	14,900	1,720	7,300	1,010	1,010	41	2,010	1,300
Birch, yellow	0.66	43	7.2	9.2	16,600	2,010	8,170	970	920	55	1,880	1,260
Cherry, black	0.53	35	3.7	7.1	12,300	1,490	7,110	690	560	29	1,700	950
Cottonwood, eastern	0.43	28	3.9	9.2	8,500	1,370	4,910	370	580	20	930	430
Elm, American	0.55	35	4.2	9.5	11,800	1,340	5,520	690	660	39	1,510	830
Elm, rock	0.66	44	4.8	8.1	14,800	1,540	7.050	1,230		56	1,920	1,320
Sweetgum	0.55	36	5.4	10.2	12,500	1,640	6,320	620	760	32	1,600	850
Hickory, shagbark	0.77	50	7.0	10.5	20,200	2,160	9,210	1,760		67	2,430	
Mahogany‡ (Swietenia spp)	0.51	34	3.5	4.8	11,460	1,500	6,800	1,100	750		1,230	800
Maple, sugar	0.68	44	4.9	9.5	15,800	1,830	7,830	1,470		39	2,330	1,450

Oak, red, northern	0.66	44	4.0	8.2	14,300	1,820	6,760	1,010	800	43	1,780	1,290
Oak, white	0.71	48	5.3	9.0	15,200	1,780	7,440	1,070	800	37	2,000	1,360
Poplar, yellow	0.45	29	4.2	7.6	10,100	1,580	5,540	500	540	24	1,190	540
Tupelo, black	0.55	35	4.4	7.7	9,600	1,200	5,520	930	500	22	1,340	810
Walnut, black	0.56	38	5.2	7.1	14,600	1,680	7,580	1,010	690	34	1,370	1,010
Softwoods												
Cedar, western red	0.34	23	2.4	5.0	7,500	1,110	4,560	460	220	17	860	350
Cypress	0.48	32	3.8	6.2	10,600	1,440	6,360	780	270	24	1,000	510
Douglas fir, coast	0.51	34	4.8	7.6	12,400	1,950	7,240	800	340	31	1,160	710
Hemlock, eastern	0.43	28	3.0	6.8	8,900	1,200	5,410	650		21	1,060	500
Hemlock, western	0.44	29	4.3	7.9	11,300	1,640	7,110	550	340	26	1,250	540
Larch, western	0.59	38	4.5	9.1	13,100	1,870	7,640	980	430	35	1,360	830
Pine, red	0.47	31	3.8	7.2	11,000	1,630	6,070	600	460	26	1,210	560
Pine, ponderosa	0.42	28	3.9	6.3	9,400	1,290	5,320	580	420	19	1,130	460
Pine, eastern white	0.37	24	2.1	6.1	8,600	1,240	4,800	440	310	18	900	380
Pine, western white	0.42	27	2.6	5.3	9,700	1,460	5,040	470		23	1,040	420
Pine, shortleaf	0.54	36	4.4	7.7	13,100	1,760	7,270	820	470	33	1,390	690
Redwood	0.42	28	2.6	4.4	10,000	1,340	6,150	700	240	19	940	480
Spruce, sitka	0.42	28	4.3	7.5	10,200	1,570	5,610	580	370	25	1,150	510
Spruce, white	0.45	28	4.7	8.2	9,800	1,340	5,470	460	360	20	1,080	480

*Tabulated from "Wood Handbook," *Tropical Woods* no. 95, and unpublished data from the U.S. Forest Products Laboratory.
†Tensile strength parallel to grain to be taken as equal to modulus of rupture in bending.
‡Central American *Swietenia* ssp.

Baumeister T: Marks' Standard Handbook For Mechanical Engineers, 8th ed. New York, McGraw-Hill, p 6-124, 1978. Reprinted by Permission.

Table 7-12: Refrigerant Applications by Appliance and by Type of Refrigerant

Appliance	Refrigerant
Domestic refrigerator	R-12, R-22
Domestic food freezers	R-12, R-22, R-502
Automobile air conditioning	R-12
Cryogenic applications	R-13, R-503
Home air conditioning	R-22, R-500
Public building air conditioning:	
Low capacity	R-12, R-22
Medium capacity	R-11, R-12, R-22
High capacity	R-11, R-12
Ship-board air conditioning	R-11, R-12, R-22
Frozen food delivery service	R-22, Solid carbon dioxide
Metal shrinking	Nitrogen
Industrial process	R-11
Food freezants	Carbon dioxide, nitrogen
Cleaning, absorbing moisture	R-11

Refrigerant	Compressor Type	Application
R-11	Centrifugal	Large air conditioning systems ranging from 200 to 2000 tons in capacity. Refrigerating systems for industrial process water and brine.
R-12	Reciprocating Centrifugal Rotary	Large air conditioning and refrigeration systems. Small household refrigerators including frozen food and ice cream cabinets, food locker plants, water coolers, room and window air conditioners and others. Principal refrigerant in automobile air conditioning.
R-22	Reciprocating Centrifugal	Residential and commercial air conditioning. Food-freezing plants, frozen-food storage and display cases and many other medium and low-temperature applications.
R-500	Reciprocating	Small home and commercial air conditioning equipment and in household refrigeration—especially in areas where 50 cycle current is common.
R-502	Reciprocating	Frozen food and ice cream display cases, warehouses and food freezing plants. Medium-temperature display cases, truck refrigeration and heat pumps.
R-503	Reciprocating	Low-temperature systems down to about −130 F. (−90 C.).
R-13	Reciprocating	Low-temperature systems down to about −130 F. (−90 C.) in cascade systems.
R-113	Centrifugal	Small to medium air conditioning systems and industrial cooling.

Althouse: Modern Refrigeration and Air Conditioning, Goodheart-Willcox Pub., p 298, 1979. Reprinted by Permission.

Table 7-13: Thermal Conductivities of Insulating Materials for High Temperature

Material	Bulk density, lb per cu ft	Max temp, ° F	100° F	300° F	500° F	1000° F	1500° F	2000° F
Asbestos paper, laminated	22.	400	0.038	0.042				
Asbestos paper, corrugated	16.	300	0.031	0.042				
Diatomaceous earth, silica, powder	18.7	1500	0.037	0.045	0.053	0.074		
Diatomaceous earth, asbestos and bonding material	18.	1600	0.045	0.049	0.053	0.065		
Fiberglas block, PF612	2.5	500	0.023	0.039				
Fiberglas block, PF614	4.25	500	0.021	0.033				
Fiberglas block, PF617	9.	500	0.020	0.033				
Fiberglas, metal mesh blanket, #900	—	1000	0.020	0.030	0.040			
Glass blocks, average values	14-24	1600	—	0.046	0.053	0.074		
Hydrous calcium silicate, "Kaylo"	11.	1200	0.032	0.038	0.045			
85% magnesia	12.	600	0.029	0.035				
Micro-quartz fiber, blanket	3.	3000	0.021	0.028	0.042	0.075	0.108	0.142
Potassium titanate, fibers	71.5	—	—	0.022	0.024	0.030		
Rock wool, loose	8-12	—	0.027	0.038	0.049	0.078		
Zirconia grain	113.	3000	—	—	0.108	0.129	0.163	0.217

Baumeister T: Marks' Standard Handbook For Mechanical Engineers, 8th ed. New York, McGraw-Hill, p 4-65, 1978. Reprinted by Permission.

Table 7-14: Performance of Adhesive Resins (Rating 1 = poorest or lowest, 10 = best or highest)

	Adherence to					Resistance to				
Adhesive resin	Paper	Wood	Metal	Ceramics	Rubbers	Water	Solvents	Alkali	Acids	Relative cost
Alkyd	6	7	5	6	7	7	2	2	5	3
Cellulose acetate	4	3	1	3	5	2	3	1	3	6
Cellulose acetate butyrate	3	3	1	4	5	2	3	1	3	7
Cellulose nitrate	5	5	1	5	5	3	2	2	4	5
Ethyl cellulose	3	3	1	3	5	2	3	3	3	6
Methyl cellulose	5	1	1	3	3	1	6	3	3	6
Carboxy methyl cellulose	6	1	2	3	2	1	6	1	4	7
Epoxy resin	10	10	8	8	8	8	9	9	8	7
Furane resin	8	7	1	8	7	8	10	10	8	5
Melamine resin	10	10	2	2	2	7	9	5	5	5
Phenolic resins	9	8	2	6	7	8	10	7	8	4
Polyester, unsat'd	6	8	2	5	7	7	6	1	6	4
Polyethylacrylate	3	4	3	5	6	8	2	5	7	5
Polymethylmethacrylate	2	3	2	3	6	8	3	8	7	5
Polystyrene	1	3	2	2	5	8	1	10	8	3
Polyvinylacetate	8	7	7	7	3	3	3	4	6	3
Polyvinyl alcohol	6	2	2	4	6	1	7	1	3	6
Polyvinyl acetal	5	7	8	7	7	8	5	3	5	7
Polyvinyl chloride	5	7	6	7	6	8	6	10	9	4
Polyvinyl acetate chloride	6	8	6	7	5	8	5	9	9	4

Polyvinylidene copolymer	4	7	6	7	7	8	7	8	9	4
Silicone T.S.	4	6	7	7	8	10	7	6	6	6
Urethane T.S.	8	10	10	9	10	7	8	4	4	9
Acrylonitrile rubber	3	6	8	6	9	7	5	8	8	5
Polybutene rubber	3	3	6	2	8	8	3	10	9	5
Chlorinated rubber	3	5	7	4	7	6	3	10	9	5
Styrene rubber	5	7	6	5	8	7	3	10	9	4

Baumeister T: Marks' Standard Handbook For Mechanical Engineers, 8th ed. New York, McGraw-Hill, p 6-142, 1978. Reprinted by Permission.

Table 7-15: Properties and Uses of Various Adhesives

Basic type	Trade name or designation	No. of components	Curing cycle, time at temp	Service temp range, °C	Adherends	Main uses	Remarks
				Animal			
Animal (hide)	(64) Jelly Glue	1	Melted at 70-75°C. Sets on cooling		Paper, wood, textiles	Woodworking, carpet materials, paper, bookbinding	May be thinned with water
Animal (hide) + plasticizers	(2) Flexhide	1	Applied as a melt at 60°C	Less than 60	Paper, cellulosic materials	Bookbinding, stationery applications	Cures to permanent flexible film
Fish glue	(120) Samsom liquid fish glue	1	1 h at 20°C	60	Wood, chipboard, paper	General-purpose for porous materials	Rapid setting. Good flexibility. Moderate resistance to water. High tack.
Casein	(34) Casco casein glue 1562	1	Cold setting after 20 min standing period on mixing		Timber with moisture content	Laminated timber arches and beams, plybox beams, and engineering timber work	Full bond strength developed after seasoning period of 48 h
Casein + 60% latex	(34) Casco 1060	2	Cold setting after 20 min standing period on mixing		Aluminum, wood, phenolic formaldehyde (rigid), leather, rubber	Bonding of disimilar materials to give flexible, water-resistant bond	Flexible
				Vegetable			
Dextrine	(134) Lion Glue	1	Air drying		Paper, cardboard, leather, wood, pottery	General-purpose glue for absorbent materials	Medium drying period of 2-3 h
Dextrine-starch blend	(211) Swift K1552	1	Applied above 15°C air drying	48	Cellulosic materials, cardboard, paper	Labeling, carton sealing, spiral-tube winding	Fast setting. May be diluted with water
Gum arabic	(196A) Sello-Bond gum	1	Cold setting		Paper, cardboard	Stationery uses	Fast drying

Mineral							
Silicate	(15) Fibrous adhesive 81-27	1	8 h at 20°C	10-430	Asbestos, magnesia	Lagging asbestos cloth on high-temperature insulation	Unsuitable where moisture: not recommended for glass or painted surfaces
Silicate with china-clay filler	(48) Autostic	1	Dried at 80°C before exposure to heat	−180 to 1500	Asbestos, ceramics, brickwork, glass, silver, aluminum, steel (mild)-steel	General-purpose cement for bonding refractory materials and metals. Furnace repairs and gastight jointing of pipe work. Heat-insulating materials	Resistant to oil, gasoline, and weak acids
Animal							
Sodium silicate	(67) Pyramid No. 1	1	Dried at 20-80°C before exposure to heat	0-850	Aluminum (foil), paper, wood-wood	Fabrication of corrugated fiber board. Wood bonding, metal foil to paper lamination	Suitable for glass-to-stone bonding
Aluminum phosphate + silica filler	(97) Brimor U527	1	Dried ½ h at 20°C, then ½ h at 70°C + ½ h at 100°C + 1 hr at 200°C + 1 hr at 250°C. Repeat for 2 overcoatings and finally cure 1 hr at 350°C	750	Steels (low-alloy), iron, brass, titanium, copper, aluminum	Strain-gage attachment to heat-resistant metals. Heater-element bonding	Particularly suited to heat-resistant steels where surface oxidation of metal at high temperatures is less detrimental to adhesion
Bitumen/Latex emulsion	(29) Colset 185	1	Dried in air to a tacky state	0-66	Cork, Polystyrene (foam), polyvinyl chloride, concrete, asbestos	Lightweight thermal-insulation boards, and preformed sections to porous and nonporous surfaces. Building applications	Not recommended for constructions operated below 0°C

Table 7-15: Properties and Uses of Various Adhesives (cont'd.)

Basic type	Trade name or designation	No. of components	Curing cycle, time at temp	Service temp range, °C	Adherends	Main uses	Remarks
				Elastomers			
Natural rubber	(40) BTR Cement 41872/3	1	Air-dried 20 min at 20°C and heat-cured 5 min at 140°C		Rubber (styrene butadiene), rubber (latex), aluminum, cardboard, leather, cotton	Vulcanizing cement for rubber bonding to textiles and rubbers	May be thinned with toluene
Natural rubber in hydrocarbon solvent	(40) Silverlock 5054	1	Air-dried 10 min at 20°C and heat-cured for 20 min at 150°C	100	Hair (keratin), bristle, polyamide fiber	Brush-setting cement for natural- and synthetic-fiber materials	Resistant to solvents employed in oil, paint and varnish industries. Can be nailed without splitting
Rubber latex	(77) Dunlop A 1020	1	Air drying within 15 min		Canvas, paper, fabrics, cellulosic materials	Bonding textiles, papers, packaging materials. Carpet bonding	Resistant to heat. Should be protected from frosts, oils
Chlorinated rubber in hydrocarbon solvents	(43) Bexol 1528	1	Air-dried 10 min at 20°C and contact bonded	−20 to 60	Polyvinyl chloride, acrylonitrile butadiene styrene, polystyrene, rubber, wood	General-purpose contact adhesive	Resistant to aging, water, oils, petroleum
Styrene-butadiene rubber lattices	(77) Dunlop SF	1	Air drying		Polystyrene (foam), wood, hardboard, asbestos, brickwork	Bonding polystyrene foams to porous surface	
Neoprene/nitrile rubbers in solvent	(40) Cement 24876	1	Dried 30 min in air and bonded under pressure while tacky		Wood, linoleum, leather, paper, metals, nitrile rubbers, glass, fabrics	Cement for bonding synthetic rubbers to metals, woods, fabrics	May be thinned with ketones
Acrylonitrile rubber + phenolic resin	(150) Scotchweld AF 30 with EC 1459 primer (10% w/w solids)	1	Primer air-dried 60 min at 20°C film cured 60 min at 175°C under pressure. Pressure released on cooling at 50°C	−40 to 130	Aluminum (alloy)-aluminum to DTD 746	Metal bonding for structural applications at elevated temperatures	Subject to creep at 150°C for sustained loading

Polysulfide rubber in ketone solvent and catalyst	(35) Bosco-frene 2115-5	2	3 days at 25°C	−50 to 130, withstands higher temps. for short periods	Metals	Sealant for fuel tanks and pressurized cabins in aircraft, where good weatherproof and waterproof properties are required	Resistant to gasoline, oil, hydraulic fluids, ester lubricants. Moderate resistance to acids and alkalies
Silicone rubber	(76) Silastic 735 RTV	1	24 h at 20°C (20% R.H.). Full cure in 5 days	−65 to 260	Aluminum, titanium, steel (stainless), glass, cork, silicone rubber, cured rubber-aluminum, cured rubber-titanium, cured rubber-steel (stainless), aluminum-aluminum (2024 Alclad), cork-cork (phenolic bonded)	General-purpose bonding and sealing applications. Adhesive/sealant for situations where material is expected to support considerable suspended weight. High-pressure exposure conditions	Resistant to weathering and moisture
Reclaim Rubber	(40) Cement 21401	1	Contact bonded when tacky		Fabric, leather, wood, glass, metals (primed)	General industrial adhesive for rubber, fabric, leather, porous materials	May be thinned with toluene
Polychloroprene	(77) Dunlop RF	1	Air-dried 10-20 min at 20°C		Rubber, steel, wood, concrete	Bonding all types of rubber flooring to metals, woods, and masonry	Good heat resistance
Modified polyurethane	(35) Foaming adhesive 2065	2	3 h at 18°C to 16 h at −15°C	−80 to 110	Concrete, plaster, ceramics, glass, hardboards, wood, polyurethane (foam), phenol formaldehyde (foam), polystyrene (foam), copper, lead, steel, aluminum	Bonding rigid and semi-rigid panels to irregular wall surfaces, wall cladding and floor laying. Building industry applications	Foam remains flexible on aging even at elevated temperatures. Will withstand a 12% movement
Thermoplastic							
Nitrocellulose in ester solvent	(120) Samsom C110	1	Heat set 1 h at 60°C after wet bonding	60	Paper, leather, textiles, silicon carbide, metals	Labeling, general bonding of inorganic materials including metals	Good resistance to mineral oils

Table 7-15: Properties and Uses of Various Adhesives (cont'd.)

Basic type	Trade name or designation	No. of components	Curing cycle, time at temp	Service temp range, °C	Adherends	Main uses	Remarks
				Thermoplastic			
Modified methyl cellulose	(110) Methofas SA	1	Dries in air		Vinyl-coated paper, polystyrene foam	Heavy-duty adhesive. Decorating paper and plastics	Contains fungicide to prevent biodeterioration
Ethylene vinyl acetate copolymer + resins	(83) Evo-Stik Therma-flow RDA 6876	1	Film transfer at 70-80°C followed by bonding at 150-160°C	60 or 1 h at 90	Cotton (duck)-cotton, resin rubber-leather, melamine laminate—plywood, steel (mild)-steel, acrylic (sheet) acrylic	Metals, laminated plastics, and textiles. Fabrication of leather goods. Lamination work	Good electrical insulation
				Animal			
Polyvinyl acetate	(120) Samson 8100 Resimul		Rapid setting		Paper, Cardboard	Carton sealing in packaging industry	Resistant to water
Synthetic polymer blend	(80) Eastobond M3	1	Applied as a melt at 177°C	71	Paper, cardboard, polythene (coated materials)	Carton and paper-bag sealing. Packaging	
Polychloroprene/resin blend in solvent	(63) Cow PA 74	2	Air dried 10 min at 20°C and cured 4 days at 20°C to 7 h at 75°C		Chlorosulfonated polythene, Polychloroprene fabrics, polyamide fabrics, leather, wood, textiles	Bonding synthetic rubbers and porous materials. Primer for polyamide-coated fabrics such as nylon, terylene	
Polychloroprene	(77) Dunlop RF	1	Air dried 10-20 min at 20°C		Rubber, steel, wood, concrete	Bonding all types of rubber flooring to metals, woods, and masonry	Good heat resistance

Saturated polyester + isocyanate catalyst in ethyl acetate	(45) Herberts EP adhesive lacquer	2	Solvent evaporation and press cured at 40-80°C when tacky		Cellulose, cellulose acetate, polyolefins (treated film), polyvinyl chloride (rigid), paper, aluminum (foil), copper (foil)	Lamination of plastic films to themselves and metal foils for packaging industry, printed circuits	Resistant to heat, moisture, and many solvents
Cyanoacrylate *(anaerobic)*	(121) I.S.-12	1	15 s to 10 min at 20°C substrate-dependent	Melts at 165	Steel-steel, steel-aluminum, aluminum-aluminum, butyl rubber-phenolic	Rapid assembly of metal, glass, plastics, rubber components	*Anaerobic* adhesive. Curing action is based on the rapid polymerization of the monomer under the influence of basic catalysts. Absorbed water layer on most surfaces suffices to initiate polymerization and brings about bonding
Polyacrylate resin *(anaerobic)*	(121) Loctite 308 impact	1	3 min at 120°C to 45 min at 65°C or 7 days at 20°C	−55 to 95	Aluminum-aluminum	Assembly requirements requiring high resistance to impact or shock loading. Metals, glass, and thermosetting plastics	*Anaerobic* adhesive
Thermosetting							
Urea formaldehyde	(34) Cascamite "one shot"	1	9 h at 10°C to 1 h at 21°C after mixing powder with water (22%)		Wood, phenolic laminate	Wood gluing and bonding on plastic laminates to wood. Plywood, chipboard manufacture. Boat building and timber engineering	Excess glue may be removed with soapy water
Phenolic formaldehyde + catalyst PX-2Z	(34) Cascophen PC-1	2	Cold setting		Wood	Timber and similar porous materials for outdoor-exposure conditions. Shop fascia panels	Good resistance to weathering and biodeterioration

Table 7-15: Properties and Uses of Various Adhesives (cont'd.)

Basic type	Trade name or designation	No. of components	Curing cycle, time at temp	Service temp range, °C	Adherends	Main uses	Remarks
				Thermosetting			
Resorcinol formaldehyde + catalyst RXS-8	(34) Cascophen RS-216-M	2	Cured at 16°C to 80°C under pressure		Wood, asbestos, aluminum, phenolic laminate, polystyrene (foam), polyvinyl chloride, polyamide (rigid)	Constructional laminates for marine craft. Building and timber applications. Aluminum-plywood bonding. Laminated plastics	Recommended for severe outdoor-exposure conditions
Epoxy resin + catalyst	(150) Scotchweld EC1614-B/A	2	24-48 h at 20°C to 20 min at 120°C	100	Steel, glass, polyester-glass fiber composite, aluminum-aluminum	General-purpose structural adhesive	
Epoxy resin + catalyst	(150) Scotchweld EC1838 B/A	2	8 h at 24°C to 2 hr at 66°C to 45 min at 121°C	65	Steel, copper, zinc, silicon carbide, wood, masonry, polyester-glass fiber composite, aluminum-aluminum	Bonding of metals, glass, ceramics, and plastic composites	Cures to strong, durable bond
Epoxy + steel filler (80% w/w)	(73) Devcon A	2	1-2 h at 21°C	120	Iron, steel, aluminum, wood, concrete, ceramics, aluminum-aluminum	Industrial maintenance repairs. Metallic tanks, pipes, valves, engine casings, castings	Good resistance to chemicals, oils, water
Epoxy + amine catalyst (ancamine LT)	(198) Epikote 828	2	2-7 days at 20°C for 33% w/w catalyst content	−5 to 60	Concrete, stonework	Repair of concrete roads and stone surfaces	Excellent pigment-wetting properties. Effective under water and suited to applications under adverse wet or cold conditions
Epoxy resin (modified)	(165) Bondmaster M620	1	4-5 h at 149°C to 20 min at 230°C to 7 min at 280°C	150	Aluminum, steel, ceramics	One-part structural adhesive for high-temperature applications	Good gap-filling properties for poorly fitting joints. Resistant to weather, galvanic action

Epoxy	(61) Mereco X-305	2	45 s at 20°C		Gem stones, glass, steel, aluminum-aluminum	Rapid assembly of electronic components, instrument parts, printed circuits. Stone setting in jewelry, and as an alternative to soldering	
Epoxy resin in solvent + catalyst	(225D) BR-600	3	8 h at 52°C to ½ h at 121°C	−270 to 371	Aluminum and magnesium alloys for elevated temperature service	Strain gages for cryogenic and elevated-temperature use. Micro measurement	Cured material resists outgassing in high vacuum
Animal							
Epoxy polyamide	(13) DEK epoxy	2	8 h at 20°C to 15 min at 100°C	100	Copper, lead, concrete, glass, wood, fiberglas, steel-steel, aluminum-aluminum	Metals, ceramics, and plastics bonding. Building and civil engineering applications	Resists water, acids, oils, greases
Epoxy/polysulfide	(57) Araldite AY111 + HY111	2	24 h at 20°C to 3 h at 60°C to 20 min at 100°C		Asbestos (rigid), ceramics, glass-fiber composites, carbon, polytetrafluoroethylene (treated), polyester (film), polystyrene (treated), copper (treated), tungsten carbide, magnesium alloys, aluminum-aluminum, steel (stainless)-steel	Cold-setting adhesive especially suitable for bonding materials with differing expansion properties	Cures to flexible material. Resistant to water, petroleum, alkalies, and mild acids
Phenol furfural + acid catalyst	(34) Cascote 1376	2	2 days at 21°C		Alumina, carbon (graphite)	Formulation of chemically resistant cements. Bedding and joining chemically resistant ceramic tiles	Extremely resistant to abrasion and heat

Table 7-15: Properties and Uses of Various Adhesives (cont'd.)

Basic type	Trade name or designation	No. of components	Curing cycle, time at temp	Service temp range, °C	Adherends	Main uses	Remarks
Pressure-sensitive							
	Fluoro-Plastics, Inc. Adhesive 80		Heated by air drying for several hours or 15-30 min at 210°C		Teflon-Teflon, Teflon-metal		Good resistance to acids and alkalies. Excellent electrical properties
Miscellaneous							
Ceramic-based	(62A) Denex-2 ceramic cement	1	Dried for ½ h at 77°C and cured ½ h at 100°C + 1 h at 200°C + 1 h at 250°C. Postcured, 1 h at 350°F	816	Metals	Strain gages, temperature sensors for elevated-temperature work	

Baumeister T: Marks' Standard Handbook For Mechanical Engineers, 8th ed. New York, McGraw-Hill, p 6-143 through 6-148, 1978. Reprinted by Permission.

Table 7-16: Average Wattage and Current Draw Ratings of Appliances

Appliance	Average Rated Watts	Average Current Draw (amps)*
Air Cleaner	50	0.42
Air Conditioner (Room)	1,000	8.33
Bed Covering	175	1.46
Blender (Food)	300	2.50
Broiler	1,200	10.00
Carving Knife	90	0.75
Clock	2	0.02
Clothes Dryer	4,800	40.00
Coffee Maker	875	7.29
Deep Fat Fryer	1,425	11.88
Dehumidifier	250	2.08
Dishwasher	1,200	10.00
Egg Cooker	510	4.25
Electrostatic Cleaner	60	0.50
Fan (Attic)	375	3.13
Fan (Circulating)	90	0.75
Fan (Rollaway)	185	1.54
Fan (Window)	195	1.63
Floor Polisher	310	2.58
Food Freezer	300	2.50
Food Mixer	125	1.04
Food Disposal	400	3.33
Fruit Juicer	100	0.83
Frying Pan	1,185	9.88
Germicidal Lamp	20	0.17
Hair Dryer	300-1500	2.50-12.50
Heat Lamp (Infrared)	250	2.08
Heat Pump	9,500	79.17
Heater (Radiant)	1,300	10.83
Heating Pad	65	0.54
Hot Plate	1,225	10.21
Humidifier	65-175	0.54-1.46
Incinerator	600	5.00
Iron (Hand)	1,100	9.17
Ironer (Mangle)	1,500	12.50
Microwave Oven	1,450	12.08
Oil Burner	250	2.08
Radio	70	0.58
Radio/Phono Combo	110	0.92
Range:		
With Oven	11,850	98.75
With Self-Clean Oven	12,200	101.67
Refrigerator	240	2.00
Refrigerator/Freezer	330	2.75
Roaster	1,340	11.17
Sandwich Grill	1,160	9.67
Sewing Machine	75	0.63
Shaver	15	0.13
Sunlamp	285	2.38

Table 7-16: Average Wattage and Current Draw Ratings of Appliances (cont'd.)

Appliance	Average Rated Watts	Average Current Draw (amps)*
Television:		
Black & White:		
Tube Type	100	0.83
Solid State	50	0.42
Color:		
Tube Type	240	2.00
Solid State	145	1.21
Toaster	1,125	9.38
Trash Compactor	400	3.33
Tooth Brush	1	0.0083
Vacuum Cleaner	600	5.00
Vibrator	40	0.33
Waffle Iron	1,200	10.00
Washing Machine (Automatic)	400	3.33
Washing Machine (Non-Automatic)	285	2.38
Waste Dispenser	450	3.75
Water Heater (Standard)	3,000	25.00
Water Heater (Quick Recovery)	4,475	37.29
Water Pump	325	2.71

*Based on 120 VAC, 60 Cy., excluding starting current draw, where applicable. When referencing this chart one should consider factors such as the actual size of the applicance under investigation and the individual usage. Appliances with motors may have variances of current due to power factor. It is suggested that persons trained in electronics be contacted when an investigation involves electrical appliances.

Table 7-17: Electric Lamps

Lamp size	Approx. surface temp. (°F)	Approx. filament temp. (°F)
40-100 Watts	200	4550
150-200 Watts	295	4726
300-500 Watts	380	4830
250-Watt Heat Lamp	400	

Table 7-18: Carbonization

230 °F	Water and aromatics are dispelled
320 °F	Carbon monoxide, carbon dioxide, and resins are dispelled
446 °F	Browning (carbonization) begins
504 °F	Pyrophoric carbon is formed
572 °F	Charcoal is formed

Table 7-19: Temperature Conversion

°F = °C times 1.8 plus 32

°C = °F minus 32 divided by 1.8

Table 7-20: Predominant Flame Colors of Materials

Element	Color	Element	Color
Lithium	Deep red	Antimony	Blue-green
Strontium	Crimson	Copper	Green-blue
Calcium	Yellow-red	Arsenic	Light blue
Sodium	Bright yellow	Lead	Light blue
Barium	Yellow-green	Selenium	Blue
Molybdenum	Green-yellow	Indium	Deep blue
Zinc	Light green	Potassium	Purple-red
Boron	Green	Rubidium	Violet
Tellurium	Deep green	Cesium	Bluish purple
Thallium	Greenish blue		

Brady: Materials Handbook: an Encyclopedia for Purchase Agents, Engineers and Foremen, 7th ed. New York, McGraw-Hill, p 826, 1951. Reprinted by Permission.

Table 7-21: Reflecting Powers of Various Metal Surfaces

	White light directly reflected, per cent	Color Silver = 0	
Silver	90		0
Chromium	61	Blue-green	12 units
Nickel	50	Red	16 units
Stainless steel	49	Blue-green	3 units
White bronze speculum	70	Red	1 unit

Brady: Materials Handbook: an Encyclopedia for Purchase Agents, Engineers and Foremen, 7th ed. New York, McGraw-Hill, p 826, 1951. Reprinted by Permission.

Table 7-22: Reflecting Power of Various Colors in Paints

Color	Light reflection, %	Color	Light reflection, %
Flat white	85-89	Sky blue	58
Bone white	69-70	Light orchid	57
Canary yellow	68-72	Buff	47
Light ivory	70	Pea green	40
Aluminum	70	Tan	34
Cream	65-69	Peacock blue	34
Light green	66	Steel gray	30
Ivory	61-63	Brown	9
Peach	58-59		

Brady: Materials Handbook: an Encyclopedia for Purchase Agents, Engineers and Foremen, 7th ed. New York, McGraw-Hill, p 826, 1951. Reprinted by Permission.

Table 7-23: Chemical and Physical Properties of Odorants Usually Found in Natural and LP Gas

Chemical Description	Molecular Weight	Specific Gravity Gas (air = 1)	Specific Gravity Liquid (water = 1)	Sulfur Content by Weight	Sulfur Contribution at 1 lb/MMcf
Tertiary butyl mercaptan	90	3.14	0.799	35.5%	0.25 gr/100 cf
Isopropyl mercaptan	76	2.65	0.820	42.0%	0.295 gr/100 cf
Normal propyl mercaptan	76	2.65	0.836	42.0%	0.295 gr/100 cf
Secondary butyl mercaptan	90	3.14	0.830	35.5%	0.25 gr/100 cf
Thiophane	88	3.07	0.999	36.4%	0.255 gr/100 cf
Dimethyl sulfide	62	2.16	0.85	51.6%	0.632 gr/100 cf
Ethyl mercaptan	62	2.16	0.84	51.6%	0.362 gr/100 cf
Diethyl sulfide	90	3.14	0.84	35.5%	0.25 gr/100 cf
Methyl ethyl sulfide	76	2.65	0.84	42.0%	0.295 gr/100 cf

Carroll JR: Physical and Technical Aspects of Fire and Arson Investigation. Springfield, Illinois, Charles C. Thomas, p 185, 1979.

Table 7-24: Electrical Energy Produced by Generators or Alternators

Vehicle	Voltage	Maximum Amperes	Watts	BTU/Hr
Motorcycles	12	40	480	1,984
Autos	12	100	1,200	4,961
Buses	24	600	14,400	59,530
Recreational vehicles	12	150	1,800	7,441
Trucks	24	120	2,880	11,906

Carroll JR: Physical and Technical Aspects of Fire and Arson Investigation, Springfield, Illinois, Charles C. Thomas, p 304, 1979.

Table 7-25: Soldering Iron Tip Temperatures

25 Watt/120 Volt	750°F
40 Watt/120 Volt	900°F
80 Watt/120 Volt	900°F
120 Watt/120 Volt	900°F

Table 7-26: Melting Points of Materials Commonly Found in Fires

Alphabetical Listing		Listing in Order of Decreasing Temperatures	
Material	Temp. °F	Material	Temp. °F
Aluminum	1150	Tungsten (light bulb filament)	6100
Aluminum alloy (zinc)	728	Steel	2760
Brass, naval	1625	Iron, wrought	2750
Brass, red	1810	Steel (stainless)	2550
Brass, yellow	1710	Iron, cast	2150
Bronze	1910	Copper	1980
Copper	1980	Bronze	1910
Fuses, electrical	700	Brass, red	1810
Glass	1200	Brass, yellow	1710
Iron, cast	2150	Muntz metal	
Iron, wrought	2750	(Cu 60%, Zn 40%)	1660
Muntz metal		Silver (sterling)	1650
(Cu 60%, Zn 40%)	1660	Brass, naval	1625
"Pot Metal" (lead-based alloy)	550	Silver (coin)	1600
Silver (coin)	1600	Glass	1200
Silver solder		Aluminum	1150
(Pb 97.5%, Ag 2.5%)	580	Zinc	780
Silver (sterling)	1650	Aluminum alloy (zinc)	728
Soft solder (Sn 50%, Pb 50%)	421	Fuses, electrical	700
Soft solder (Sn 60%, Pb 40%)	374	Silver solder	
Steel	2760	(Pb 97.5%, Ag 2.5%)	580
Steel (stainless)	2550	"Pot Metal" (lead-based alloy)	550
Tin	450	Tin	450
Tungsten (light bulb filament)	6100	Soft solder (Sn 50%, Pb 50%)	421
Zinc	780	Soft solder (Sn 60%, Pb 40%)	374

Carroll JR: Physical and Technical Aspects of Fire and Arson Investigation, Springfield, Illinois, Charles C. Thomas, p 52, 1979.

Table 7-27: Characteristics Of Lubricants

Component	Lubricant	Operating Temp. °F	Flash Point °F
Power steering fluid	Oil	200	350
Front wheel bearings (disc)	Grease	300	450
Four-wheel drive, front transfer case	90 weight gear oil	200	400
Engine lubricating oil			
10W	Oil	220	400
20	Oil	220	440
30	Oil	220	460
40	Oil	220	480
50	Oil	220	500
Universal joints	Grease	220	520
Rear axle	Grease	220	520
Transmission oil			
Automatic	Oil	220	370
Manual	90 weight oil	220	400
Rear end differential	Gear oil	200	400

Carroll JR: Physical and Technical Aspects of Fire and Arson Investigation, Springfield, Illinois, Charles C. Thomas, p 307, 1979.

Table 7-28: Ignition Temperatures of Materials

Absorbant Cotton	442°F to 446°F
Acetone	1000°F
Acrylic/Methyl Methacrylate	830°F to 860°F
Acrylonitrile-butadiene-styrene	871°F (Heat-Resistant Type)
Alkyds (Paints, Lacquers)	Not Established
Aluminum	1150°F to 1500°F (Melts)
Aluminum Paint Flakes	949°F
Aropol (*See* Polyesters)	
Asphalt	905°F
Aviation Grade Gasoline	824°F (100-130 Grade) 880°F (115-145 Grade)
Bakelite (*See* Phenolic)	
Benzene (Benzol)	1044°F
Brass	1575°F to 1800°F (Melts)
Butane	761°F
Camphor	871°F
Cane Fiberboard	440°F
Carbon Soot	366°F
Castor Oil	840°F
Cellophane	468°F
Cellophane Paper	462°F
Cellulose Acetate	887°F
Cellulose Nitrate	285°F
Cellulose Nitrate Film	194°F to 279°F
Cellulose Triacetate, Fiber	1004°F
Cigarette (Avg. Surface Temp.)	550°F (No Drafting)
Cigarette (Avg. Temp. in Center)	1350°F (With Drafting)

Table 7-28: Ignition Temperatures of Materials (cont'd.)

Coal	600°F
Cologne Spirits	700°F
Corn Oil (Mazola)	740°F
Copper	1981°F (Melts)
Cotton Batting	511°F
Cotton Sheeting	464°F
Creosote	637°F
Crude Pine Gum Powder	581°F
Denatured Alcohol	750°F
Douglas Fir	489°F
Durez (*See* Phenolic)	
Epoxy	Not Established
Estane (*See* Polyurethane)	
Ethane	950°F
Ethanol	700°F
Ethyl Alcohol	793°F
Ethyl Cellulose	565°F
Filler Paper	450°F
Filter Paper	489°F
Formrez (*See* Polyesters)	
Fuel Oil, #1	490°F
Fuel Oil, #2	494°F
Fuel Oil, #3	498°F
Fuel Oil, #4	505°F
Fuel Oil, #5	Not Available
Fuel Oil, #6	765°F
Furniture Cushioning, Thermoplastic (*See* Polyurethane Rigid Foam)	
Gasoline	536°F to 853°F (Depending on Octane)
Genal (*See* Phenolic)	
Genfoam (*See* Polyurethane)	
Glass	1400°F to 1600°F (Melts)
Glazing, Thermoplastics (*See* Acrylic/Methyl Methacrylate & Polycarbonate)	
Glycerine	739°F
Hay	342°F
Heptane	452°F
Hetrofoam (*See* Polyurethane)	
Hexane	477°F
Insulation, Wire, Thermoplastic (*See* Polypropylene; Polytetrafluorethylene; Polyvinyl Chloride)	
Iron, Pure	2795°F (Melts)
Iron, Gray Cast	2246°F (Melts)
Iron, White Cast	2102°F (Melts)
Iron, Wrought	2750°F (Melts)
Iron, Cast	2000°F to 3800°F (Melts)
Isofoam (*See* Polyurethane)	
Isopropyl Alcohol (Rubbing)	852°F
Jute Fiber	379°F
Kauresin (*See* Phenolic)	
Kerosene	450°F to 490°F
Kodel (*See* Polyesters)	
Laminac (*See* Polyesters)	
Leather	414°F

Table 7-28: Ignition Temperatures of Materials (cont'd.)

Light Diffusers, Thermoplastic (*See* Acrylic/Methyl Methacrylate and Polycarbonate)	
Linseed Oil	650°F
Longleaf Pine	428°F
Magnesium	945°F
Magnesium, Cast	1144°F
Magnesium, Coarse Powder	950°F
Magnesium, Fine Powder	883°F
Magnesium, Ribbon	1004°F
Match Heads	325°F
Melamine Formaldehyde (Electrical Grade)	1114°F
Melamine Tableware	1150°F to 1190°F
Methane	999°F
Methyl Alcohol	878°F
Mondur (*See* Polyurethane)	
Mylar (*See* Polyesters)	
Newspaper	785°F
Newsprint Paper	363°F to 448°F
Niax (*See* Polyurethanes)	
Nylon (*See* Polyamide)	
Nylon-6	905°F
Nylon 6/6, Fiber	990°F
Nylon 6/10	856°F
Octane	448°F
Oiled Viscose Rayon	478°F
Olive Oil	650°F
Oregon Oak	500°F
Oxidized Linseed Oil	864°F
Paint Film	864°F
Paper Birch	399°F
Paraffin Wax	437°F
Partitions, Glass-reinforced (*See* Polyester)	
Peanut Oil	833°F
Pentane	588°F
Phenolformaldehyde (General Purpose)	919°F
Phenolformaldehyde, Paper Base Laminate, NEMA Type XX	853°F
Phenolic	1060°F to 1076°F
Pluracol (*See* Polyurethane)	
Plyophen (*See* Phenolic)	
Polyamide, (Nylon)	795°F
Polycarbonate	930°F
Polyester	810°F to 910°F
Polyfoam (*See* Polyurethane)	
Polyether	780°F
Polyethylene	660°F
Polylite (*See* Polyester)	
Polymethyl Methacrylate	832°F to 864°F
Polypropylene	756°F
Polypropylene, Fiber	1058°F
Polystrene, Beads	915°F
Polystrene, High Impact, Flame Retardant	793°F
Polystyrene, High Impact, High Rubber	892°F
Polystyrene, Medium Impact, Low Rubber	874°F
Polytetrafluorethylene	984°F to 987°F

Table 7-28: Ignition Temperatures of Materials (cont'd.)

Polyurethane Rigid Foam	780°F
Polyvinyl Chloride (Clothing, Upholstery, etc.)	780°F
Polyvinyl Chloride-Acetate	815°F to 1035°F
Polyvinyl Chloride, Semi-rigid, 105-C Appliance Compound	797°F
Polyvinylidene Chloride	992°F
Ponderosa Pine (Sapwood)	457°F
Powder Tin	842°F
Propane	871°F
Redwood	467°F
Scotchfoam (*See* Polyurethane)	
Silicone	1022°F to 1047°F
Sitka Spruce	482°F
Soy Bean Oil	833°F
Sperm Oil	586°F
Sulphur	450°F
Shellac	810°F
Steel	2350°F to 2675°F (Melts)
Stoddard Solvent	450°F
Styrene	914°F
Styrene-Acrylonitrile Copolymer	850°F
Styrene-Methyl Methacrylate Copolymer	905°F
Tan Bark Oil	448°F
Tableware, Thermosetting (*See* Melamine)	
Teflon (*See* Polytetrafluorethylene)	
Test Tubes, Thermoplastic Throwaway (*See* Cellulose Nitrate)	
Thermal Insulation, Foam, Thermoplastic (*See* Polyurethane Rigid Foam)	
Thermoplastics (*See Also* Acrylic/Methyl Methacrylate, Cellulose Nitrate, Polyamide (Nylon), Polycarbonate, Polyethylene, Polypropylene, Polystyrene, Polytetrafluorethylene (Teflon), Polyurethane, Polyvinyl Chloride)	
Thermosetting (*See Also* Alkyd, Epoxy, Melamine, Phenolic, Polyester, Silicone, Urea Formaldehyde)	
Tin	425°F to 475°F (Melts)
Tin, Coarse Powder	1094°F
Tobacco	342°F
Toluene	1026°F
Tung Oil	855°F
Turpentine	488°F
Unoiled Viscose Rayon	453°F
Valox (*See* Polyesters)	
Varnish	864°F
Viscose Rayon	546°F
Western Red Cedar	468°F
White Oak	410°F
White Pine	406°F
Wire Insulation, Thermoplastic (*See Also* Polypropylene, Polytetrafluorethylene, Polyvinyl Chloride)	
Wood Fiberboard	424°F to 444°F
Xylene-(M)	986°F
Xylene-(O)	924°F
Xylene-(P)	986°F
Zinc	750°F to 775°F (Melts)
Zinc Powder	1202°F

Index

A

B

C

D

E

F

N

O

P

R

S

T

U

V

W

Z

NOTES

NOTES

NOTES

NOTES

NOTES

NOTES

NOTES

NOTES

NOTES

NOTES

Related Titles from BRADY

Investigating the Fireground

Calvin C. Phillipps, David A. McFadden

This important text stresses a systematic search for fire cause, emphasizing that fire investigation can identify accidental causes as well as arson. Includes numerous case histories.
1982/281 pp/illus/paper/ISBN 0-89303-074-0/D 0740-3

Engine Company Fireground Operations

Hal Richman

Profusely illustrated, this text explains the engine company's operations, objectives, and organization, and describes the use of its firefighting tools. Discusses characteristics of various types of fires and explains how radiant and convected heat contribute to fire intensity and extension. All information is organized for increased accessibility compared to other books in the field.
1975/176pp/illus/paper/ISBN 0-87618-230-9/D 2309-5

Truck Company Fireground Operations

Hal Richman

This book covers the duties and operations of truck companies at working fires. Written in a clear simple style, with illustrations that vividly depict important learning points and procedures. Essential to all departments, even those which must function without separate truck companies.
1977/206pp/illus/paper/ISBN 0-87618-976-3/D 0594-5

Firefighting Operations in Garden Apartments and Townhouses

Glenn Gaines

With garden apartments and townhouses fast becoming the primary types of dwelling in the US, firefighters must be able to solve special kinds of problems. Describes construction methods and materials, factors contributing to fire spread, and accessibility. Discusses preplanning for attacking such fires, and then outlines tactics for such attacks. Outstanding illustrations and text make this book a must for all firefighters.
1978/144pp/illus/paper/ISBN 0-87618-885-4/D 8854-4

Wildfires: Prevention and Control

Harry P. Gaylor

Outlines the essential procedures for preventing and fighting wildfires in grasslands, brush, forests, and uplands. Recognizes prevention as the key to wildfire control, and explains how to coordinate prevention activities by the public with personnel training programs.
1974/336pp/illus/paper/ISBN 0-87618-131-0/D 1310-4

Marine Fire Prevention, Firefighting and Fire Safety

Robert J. Brady Co. for the National Maritime Research Center

Here's the only comprehensive handbook available to address the causes of hostile shipboard fire, discuss preventative measures and give specific yet complete instructions in mechanical and direct attack fire control and firefighting. A "must to read" for anyone in shipping and other maritime industries.
1979/388pp/illus/paper/ISBN 0-87618-994-X/D 994X-5